W9-BVY-910

Triple Repeat Diseases of the Nervous System

ADVANCES IN EXPERIMENTAL MEDICINE AND BIOLOGY

A Continuation Order Plan is available for this series. A continuation order will bring delivery of each new
volume immediately upon publication. Volumes are billed only upon actual shipment. For further information
please contact the publisher.

Triple Repeat Diseases of the Nervous System

Edited by
Lubov T. Timchenko
Department of Medicine
Baylor College of Medicine
Houston, Texas, U.S.A.

Kluwer Academic / Plenum Publishers
New York, Boston, Dordrecht, London, Moscow

Landes Bioscience / Eurekah.com
Georgetown, Texas, U.S.A.

BS

Library of Congress Cataloging-in-Publication Data

CIP applied for but not received at time of publication.

Triple Repeat Disorders of the Nervous System
Edited by Lubov T. Timchenko
ISBN 0-306-47417-4
AEMB volume number: 516

©2002 Kluwer Academic / Plenum Publishers and Landes Bioscience

Kluwer Academic / Plenum Publishers
233 Spring Street, New York, NY 10013
http://www.wkap.nl

Landes Bioscience
810 S. Church Street, Georgetown, TX 78626
http://www.landesbioscience.com; http://www.eurekah.com
Landes tracking number: 1-58706-055-8

10   9   8   7   6   5   4   3   2   1

A C.I.P. record for this book is available from the Library of Congress.

6/14/07

PREFACE

World of Unstable Mutations

The book "Triplet Repeat Diseases of the Nervous System" overviews the latest data on several disorders associated with unstable mutations. This field of research is progressing extremely fast. The number of polymorphic mutations and diseases caused by these mutations is increasing almost every month. There is a strong interest to molecular bases of triplet repeat disorders. This is explained by growing necessity to develop molecular approaches for cure of these diseases. Therefore, the authors of this book describe unstable mutations with the emphasis on molecular pathology. Broad discussion is presented on how polymorphic expansions cause cell dysfunction.

- The first chapter of the book focuses on the molecular pathological processes that originate "unstable" mutations. The authors review several available models by which normal "stable" region of DNA become pathogenic and discuss possible mechanisms causing DNA instability.
- The other chapters of the book describe inherited diseases associated with different types of unstable mutations. Based on the location of mutation in the disease gene, polymorphic expansions of the nervous system can be divided into two major groups. First group includes disorders with unstable expansions within the open reading frame of the gene such as Spinocerebellar Ataxias caused by polyglutamine expansions. The second group includes diseases caused by expansions situated within the untranslated regions of the gene.
- In Chapter 3, the authors summarize comprehensive information (clinical, genetic, and molecular) on all known polyglutamine diseases with the emphasis on the common and distinctive features between different disorders in this group.
- Recent studies suggested that some Spinocerebellar Ataxias might be caused not by polyglutamine track, but by unstable repeats in untranslated regions of DNA. The authors in Chapter 4 discuss Spinocerebella Ataxia 10 that is associated with polymorphic ATTCT repeat in the intron of unknown gene.
- The chapter 2 describes Myotonic Dystrophy 1 that is caused by unstable CTG repeats within the 3' UTR of the disease gene. Multiple hypotheses

were proposed to explain this disease. The authors analyze several mouse models for DM1 disease and discuss how CTG repeat is expanded in patients and why such expansion causes pathology.

- Chapter 5 describes another example of disorder with polymorphic mutation in non coding region of the gene—Friedreich Ataxia. In this disease, GAA triplet repeat is located in the first intron of frataxin. The author provides comprehensive information on clinical and genetic aspects of this disease and analyzes a variety of possible molecular pathways responsible for occurrence of mutation and development of pathology.

What should we expect in future in this field? Identification of novel unstable mutations will be accompanied by identification of molecular pathways by which these mutations cause a disease. Emerging information from different fields of science (protein and RNA biology, physiology, crystallography, gene therapy, etc) will help to develop strategy to prevent and cure disorders with unstable mutations.

The authors of this book are thankful to editorial coordinators and publishers for their support and tireless help.

Lubov T. Timchenko
Department of Medicine
Baylor College of Medicine
Houston, Texas, U.S.A.

PARTICIPANTS

Tetsuo Ashizawa
Department of Neurology
Baylor College of Medicine and
    Veterans Affairs Medical Center
Houston, Texas 77030
USA

Alexis Brice
INSERM U289
Institut Fédératif di Recherche des
    Neurosciences
Département de Génétique
Cytogénétique et Embryologie
Groupe Hospitalier Pitié-Salpêtriére
75651 Paris Cédex 13
France

Thomas A. Cooper
Departments of Pathology and
    Molecular and Cell Biology
Baylor College of Medicine
Houston, TX 77030
USA

Alexandra Dürr
INSERM U289
Institut Fédératif di Recherche des
    Neurosciences
Département de Génétique
Cytogénétique et Embryologie
Groupe Hospitalier Pitié-Salpêtriére
75651 Paris Cédex 13
France

Tohru Matsuura
Department of Neurology
Baylor College of Medicine and
    Veterans Affairs Medical Center
Houston, Texas 77030
USA

Darren G. Monckton
Institute of Biomedical and Life
    Sciences
The University of Glasgow
Glasgow, G12 8QQ
UK

Massimo Pandolfo
Centre Hospitalier de la Université de
    Montréal
Hopital Notre-Dame
1560 rue Sherbrooke Est
Montréal, Quebec
H2L 4M1 Canada

Pawel Parniewski
Centre for Microbiology and Virology
Polish Academy of Sciences
ul. Ludowa 106
93-232 Lódz
Poland

Pawel Staczek
Dept. of Genetics and Microorganisms
University of Lódz
ul. Banacha 12/16
90-127 Lódz
Poland

Giovanni Stevanin
INSERM U289
Institut Fédératif di Recherche des
    Neurosciences
Cytogénétique et Embryologie
Groupe Hospitalier Pitié-Salpêtriére
75651 Paris Cédex 13
France

Steve J. Tapscott
Division of Human Biology
Fred Hutchinson Research Center
Seattle, WA 98109
USA

Lubov T. Timchenko
Section of Cardiovascular Sciences,
    Departments of Medicine, and
    Molecular Physiology & Biophysics
Baylor College of Medicine
Houston, TX 77030
USA

CONTENTS

MOLECULAR MECHANISMS
OF TRS INSTABILITY

Pawel Parniewski[1] and Pawel Staczek[2]

INTRODUCTION

Microsatellites, stretches of short, tandemly repeated motifs of one to six nucleotides are very unstable and display very high polymorphism among individuals.[1-4] Of these repeats, a special class of microsatellites, trinucleotide repeat sequences (TRS) are involved in human neurodegenerative diseases.[5,6] To date, several neurological or neuromuscular hereditary human disorders—also called mental retardation diseases—have been linked to the genetic instability of the TRS. Diseases including myotonic dystrophy, Huntington's disease, Kennedy's disease, fragile X syndrome, spinocerebellar ataxias or Friedreich's ataxia result from expansion of trinucleotide sequences such as $(CTG/CAG)_n$, $(CGG/CCG)_n$, or $(GAA/TTC)_n$ present in human genome.[7]

The unstable TRS may expand and thus, depending on their localization in the chromosomes, may disturb an expression of crucial genes. The function of the majority of genes which activity is affected remains unknown. It is clear however, that common features of the diseases mentioned above result as a consequence of the expansion of the TRS. Moreover, the inheritance of such diseases cannot be explained by Mendelian genetics due to the character of expansions described as dynamic mutations.[8,9] Mutations of this type result in a length change of DNA tracts containing repeated sequences. From the clinical point of view, the increasing length of the same TRS region causes a progressive increase in expressivity of the mutation over a number of generations. Such phenomenon is termed "anticipation".[10] There is an inverse relationship between the age of onset and the size of repeat and a direct

[1]Centre for Microbiology and Virology, Polish Academy of Sciences, ul. Lodowa 106, 93-232 Lódz, Poland and [2]Department of Genetics of Microorganisms, University of Lódz, ul. Banacha 12/16, 90-127, Poland.

relationship between expansion size and disease severity. The length of expanded TRS in afflicted families may vary from as much as tens of triplet repeats (Huntington's disease) to a few thousand repeats (myotonic dystrophy).

Over a decade of an extensive study on the nature of the genetic instabilities of TRS revealed that size alterations of these tracts may be generated via different biochemical mechanisms, including replication, transcription, DNA repair and recombination. Additionally, experiments in bacteria, yeast and mammalian systems suggest that an elevated frequency of length changes (expansions and deletions) of TRS is caused by their propensity to form unusual secondary structures.[11]

It is interesting that each of the neurodegenerative disorders mentioned above shows a highly defined threshold for the TRS tract length (usually 30-40 repeats, specific for each disease) beyond which the instability of such sequences increases dramatically, leading to the massive expansions. It is not known what drives short trinucleotide tracts present in healthy individuals to reach such threshold level. One plausible explanation of the origin of TRS-related disorders assumes a slow accumulation of small-increment length changes (SILC) which eventually pushes a repeated sequence to reach the threshold level. Beyond this level, further massive length changes (MLC) seem to be inevitable. One might assume that SILC and MLC are due to the character of DNA containing triplet repeats of different lengths (below or above the threshold) that may serve as distinct substrates for cellular factors. It cannot be excluded that maintaining the number of the repeats at the safe level results from the subtle balance between the rate of expansions versus deletions.

In this Chapter we emphasize the role of the non-B-DNA conformations as a primary source of the genetic instability of TRS. Depending on the length, type of tracts or the presence of interruptions, the character of repeated sequences creates diverse opportunity for such alternative secondary DNA structures formation. Their presence may have significant impact on the course of the metabolic processes like replication, transcription, repair and recombination occurring in the given DNA region, leading in effect to the different modes of instability.

SECONDARY DNA STRUCTURES AS A SOURCE OF TRS INSTABILITY

It has been known for some time that microsatellites can differ in repeat number among individuals and influence the integrity of genetic information. Alterations in the size (insertions and deletions) of DNA are not limited to the TRS-dependent disorders, as other microsatellite instabilities are also observed in tumors from patients with hereditary nonpolyposis colorectal cancer (HNPCC).[12-16]

Several factors may contribute to the mutational dynamics of microsatellite DNA, including number of repeats, composition and length of the repeating motif, presence of interruptions within the sequence and the rate of intracellular processes such as replication, transcription, repair, or recombination.[17] Experiments in *Escherichia coli* demonstrated that such sequences may gain or loose repeats. Most

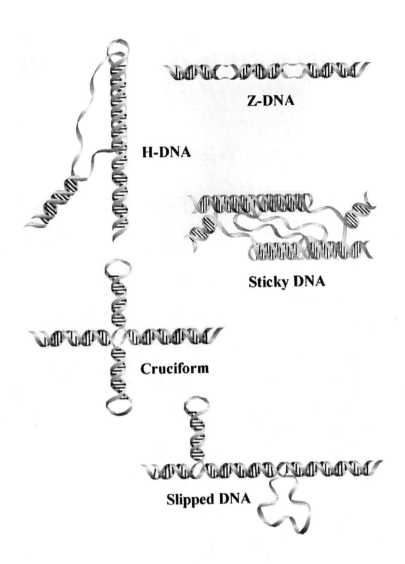

Figure 1. Non-B-DNA structures. Depending on the local base composition and the symmetry of a sequence, DNA may adopt various conformations including left-handed Z-DNA (alternating purine-pyrimidines such as $d(GC)_n$), triplexes (polypurine/polypyrimidine tracts with mirror symmetry), "sticky" DNA (association of two separate triplexes), cruciforms (arising from self-pairing inverted repeats) or slipped DNA which may form basically within any direct repeats.
Animation depicting the formation of triplex and "sticky" DNA structures is available at http://www.tamu.edu/ibt/ibtweb/stickydna.htm. Reproduced from refs. 5 and 7 with permission.

of the early work demonstrated that the instability did not depend on a RecA function of a host strain, suggesting that recombination was not the predominant mechanism generating microsatellite variability.[18] A significant feature of the direct (tandem) repeats is their intrinsic ability to form non-B-DNA conformations.[19] Unusual DNA structures (Fig. 1.1) such as left-handed Z-DNA, cruciforms, slipped-stranded DNA, triplexes, and tetraplexes may form due to their palindromic nature, under physiological conditions, also in vivo.[20-26] Such structures potentially may be hazardous for genome stability if not removed by repair mechanisms. Many experimental lines of evidence have shown that non-B-DNA-forming sequences are unstable and deleterious.

Numerous in vitro studies have demonstrated the ability of the CTG/CAG and CGG/CCG tracts to form thermodynamically stable self-complementary hairpin structures and tetraplexes.[27-29] Hairpins assembled from CTG oligomers, as revealed by NMR study, form very stable antiparallel duplexes with TT pairs, whereas CAG oligonucleotides produce much less stable conformations which are destabilized by AA mispairs.[30] This gives rise to unequal structural properties of repeated DNA during processes where single-stranded regions are involved, i.e., replication, transcription, repair or recombination. Hairpin structures will be formed and maintained more easily on the CTG strand than loops created on a strand containing CAG repeats. Similar studies confirmed that the Fragile X $(CCG/CGG)_n$ triplets could also form hairpin structures although the $(CCG)_n$ strand more readily underwent self-pairing rearrangements than the complementary $(CGG)_n$ strand.[31] In vitro measurements of the elastic constants of $(CTG/CAG)_n$ and $(CGG/CCG)_n$ and calculations of their free energy of supercoiling revealed their higher flexibility and their writhed structure in contrast with random DNA sequence.[32,33] Interestingly, CTG/CAG and CGG/CCG repeats differ in their susceptibility to nucleosome formation. While former ones are prone to bind histone proteins, the latter generally prevent formation of the nucleosomes.[34-38] However, methylation level which determines the binding constant of the histones differs significantly between short and long CGG/CCG sequences as was shown by Godde et al.[38] They pointed out that tracts shorter than 13 units upon methylation showed higher potential in nucleosome formation than long ones consisting of 74 repeats. Taken together, specific physicochemical features of these repeated sequences may be responsible for the alteration of the chromatin organization also in the neighboring regions.[39] The flexible character of these TRS and their capability to form alternative DNA structures suggest they may act as a "sink" for the accumulation of superhelical density. Superhelical tension stabilizes secondary DNA structures and may be a crucial factor promoting the formation of structural abnormalities inside long TRS motifs. Non-B-DNA conformations may influence the activity of the enzymes involved in DNA processing.

It has been shown that many DNA polymerases pause within the long stretches of the $(CTG/CAG)_n$ and $(CGG/CCG)_n$ in vitro.[40,41] The pausing sites of DNA synthesis at specific loci in the TRS depended on the length of the repeat tract, and were abolished by heating at 70°C. These results suggest that appropriate lengths of

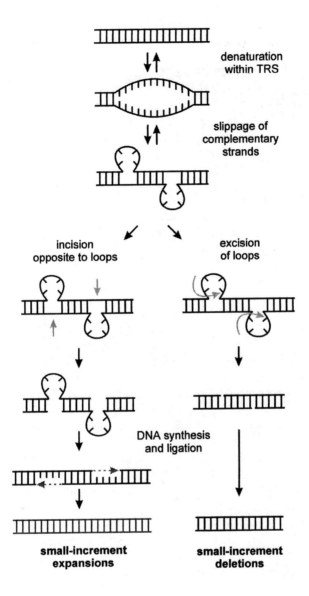

Figure 2. Mechanism for SILC. Following a denaturation and the formation of the slipped-DNA structure within the TRS, error-prone repair of resulting loop-outs produces small-increment expansions as a consequence of incisions opposite to loops and/or small-increment deletions by excision of loops.

triplet repeats adopt very stable non-B-DNA conformations that cause polymerases to pause during DNA synthesis. There is no direct evidence that polymerase stalling may produce expansions. However, such pausing may cause a primer-template realignment, which may lead to deletions and expansions, especially in repetitive sequences.

Formation of large hairpins, especially during the replication of the TRS is believed to account for massive length alterations within these repetitive sequences, including expansion events linked to neurodegenerative disorders in humans. A novel type of instability based on duplications of CTG/CAG tracts (but not CGG/CCG motifs) including neighboring sequences were reported to frequently occur when cloned in R6K plasmids.[42] Although this type of instability is not related to the neurodegenerative diseases, the presence of GAA/TTC and GAG/CTC tracts was probably responsible for the duplications in regions containing genes involved in development of neuroblastomas and malignant melanomas.[43]

Another structural aberration due to the slippage of complementary strands within the TRS is probably responsible for small deletions and expansions. Such a phenomenon was hypothesized to be the mechanism responsible for the slipped strand mispairing mutagenesis, the genetic hypermutability of dinucleotide repeat sequences in mismatch repair-deficient cells related to hereditary nonpolyposis colon cancer.[12-16,18,44,45] As shown in an in vivo E.coli model utilizing methyl-directed mismatch repair (MMR) or nucleotide excision repair (NER) defective cells, long $(CTG/CAG)_n$ motifs cloned in plasmids exhibit very frequent length changes of 1-8 repeating units.[46] These small expansions and deletions found in E.coli were studied in the absence of the repair functions since such activities would be expected to recognize and repair the looped structures formed within the slipped TRS conformation. The occurrence and the size of deletions and expansions present on plasmids isolated from single colonies were precisely monitored by the position of G to A interruptions present on the initial $(CTG/CAG)_n$ insert that served as valuable markers. The location of interruptions as compared to their original position indicated the type (expansion or deletion) and the size of the dynamic mutations observed in vivo.

Not only the $(CTG/CAG)_n$ tracts undergo dynamic mutations due to the slippage of the complementary strands. Similar analyses were also conducted on CGG/CCG fragile X and GAA/TTC Friedreich's ataxia sequences, where investigations showed that expansion and deletion products, differing in length from each other by one repeat or multiples of three base pairs could be resolved as distinct bands in polyacrylamide gel electrophoresis.[46]

Small-increment length changes (SILC) are believed to be a consequence of strand misalignment leading to a formation of single-stranded loop-outs that consist of a few repeating units (Fig. 1.2). Such bubbles may be processed by endonucleolytic activities where excision of loops gives small deletions, whereas incisions opposite to loops produce expansions. The involvement of the nucleotide excision repair in SILC remains unclear and requires further investigations. The role of methyl-directed mismatch repair in generating genetic instabilities within the TRS will be discussed in next section.

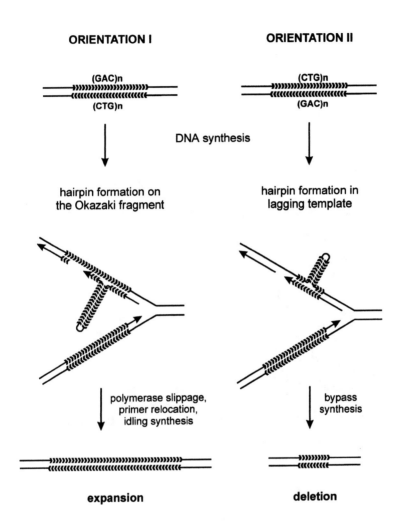

Figure 3. Mechanism of the genetic instability of the TRS during replication. Formation of the hairpin structure on the newly synthesized Okazaki fragment, primer relocation and idling synthesis will result in large expansion while the bypass synthesis through the hairpin structure formed on the template strand will produce large deletions. Note that both, deletions and expansions may also happen during leading strand synthesis but such events are substantially less frequent.

In the case of long GAA/TTC stretches of more than 59 repeats, a novel, non-B-DNA structure has been detected in supercoiled plasmids in vitro and in vivo using a bacterial model.[47] This structure is believed to originate from the self-association of two separate triplexes resulting in the formation of a very stable conformation termed

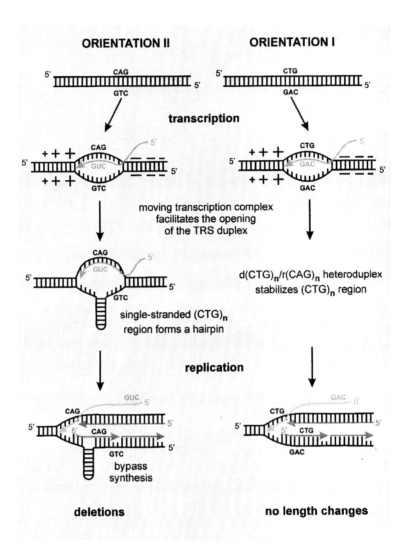

Figure 4. Correlation between replication, orientation of the (CTG/CAG)$_n$ tract, transcription, structural properties of repeated DNA and the genetic instability of the TRS.

Independently of the TRS orientation the moving transcription complex causes the opening of the DNA duplex. In orientation II (left panel) the CAG strand is being transcribed while the complementary CTG strand remains transiently single-stranded and may form stable hairpin which may be bypassed by the incoming replication complex what would lead to large deletions. In orientation I (right panel) the CTG strand is being transcribed and its interaction with the RNA polymerase complex stabilizes the CTG motif (no hairpin formation). The CAG strand remains single-stranded but cannot form stable secondary structures. Thus, the replication through the CTG/CAG tract in orientation I would not give change in number of the TRS. Note that during replication of the TRS in orientation I expansion may occur (Fig. 1.3).

"sticky DNA" (Fig.1.1). It has been shown that such structures may inhibit the process of DNA transcription and therefore could be responsible for the decreased amount of frataxin in FRDA patients.[48,49] It does not explain though how GAA tracts may expand to reach the number of required repeats.

In the next sections we will discuss more specifically the possible influence of the secondary structures on the course of major DNA metabolism pathways.

Replication

Structural propensities of repeated DNA motifs, including the TRS may cause such sequences to form slipped-stranded structures and hairpins during movement of the polymerase. In 1995 Kang et al for the first time set up the in vivo (bacterial) system in which they were able to demonstrate that long $(CTG/CAG)_n$ tracts contained on ColE1 plasmids do undergo length changes after a number of cell generations.[50] Although, the instability pattern in *E.coli* shows strong bias toward deletions, this system provided also the evidence for expansion events. The great significance of this discovery was that it proved the usefulness of the bacterial system to study the genetic instability of human repeated sequences and enabled detailed analyses of the mechanisms of expansions that are the cause of many hereditary disorders. The observed instability strongly depended on the orientation of the insert relative to the origin of replication and the length of $(CTG/CAG)_n$ motif.[51] A widely accepted model (Fig. 1.3) proposes that deletions occur if the hairpin is formed on the lagging strand template (so called orientation II). A single stranded CTG region within the replication fork forms a thermodynamically stable hairpin, which is then bypassed by incoming DNA polymerase, that in turn produces deletions. Conversely, expansions arise as a consequence of secondary DNA structures being formed during lagging strand synthesis (orientation I). Slippage of the newly synthesized repeated DNA, the formation of CTG hairpins on the Okazaki fragments, realignment of the primer, and the idle synthesis of DNA polymerase result in large expansions.[50-52] Hairpins are believed to be favored on the lagging strand, but they could also occur on the leading strand. In studies employing the single stranded bacteriophage replication model, $(CTG/CAG)_n$, $(CGG/CCG)_n$, and $(GAA/TTC)_n$ repeats underwent deletions during leading strand synthesis.[53] Interestingly, of all ten possible triplet repeats, CTG motifs are expanded a few times more frequently than the other ones.[51] A variety of studies have confirmed a dramatic influence of replication on the genetic instability of TRS in bacteria and yeast.[54-56]

As expected, the instability of both deletion and expansion events, was strongly affected by length of repeated sequences. Inserts of less than 20 CTG units, that are much less prone to form secondary structures, do not delete nor expand, while tracts consisting of approximately 50 units become unstable. This resembles the situation observed during the development of the disease in humans.

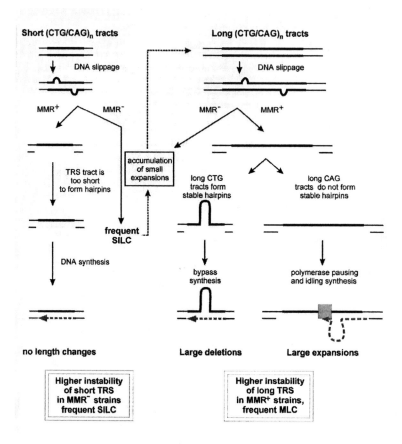

Figure 5. The effect of methyl-directed mismatch repair on the genetic stability of short and long CTG/CAG tracts in generating SILC and MLC (see text for details).

Transcription

All TRS related to human diseases are actively transcribed. Unwinding of the double-stranded DNA by moving RNA polymerase complex introduces locally high torsional stress which leads to the formation of twin domains of differential DNA supercoiling, with the regions ahead and behind the polymerase having increased positive and negative supercoiling, respectively.[57] The energy of negative superhelical turns may facilitate formation of unusual DNA structures from sequences with high propensities to undergo such a transition. Notably, it was shown that transcription could promote hairpin formation within repeating sequences in *E.coli* and formation of such structures in TRS during transcription could lead to length changes of the repeat tract. Studies in *E.coli* have shown also that transcription has large impact upon the genetic stability of $(CTG/CAG)_{175}$.[58, 59] Multiple recultivations of strains harboring TRS containing plasmids led to significant reduction of the full-length, non-deleted repeats under conditions where transcription through the repeat was induced. Similarly, transcription was found to destabilize dinucleotide repeats in yeast.[60] Transcription was also reported to be crucial in affecting the genetic instability of long CTG/CAG motifs by NER pathway in *E.coli*.[59]

A possible correlation between replication, orientation of the TRS, active transcription, structural properties of repeated DNA and the genetic instability of TRS is shown in Figure 1.4. The top strand of the duplex TRS on both sides of the figure serves as the transcribed strand, as well as the leading strand template for DNA synthesis. The left side of Figure 1.4 represents orientation II (CTG strand serves as the lagging strand template), whereas the right side shows orientation I (CTG is within the Okazaki fragment). Transcription of the CAG strand leads to deletions, whereas transcription of the CTG strand elicits a much lower frequency of deletions. The model proposes that as the CAG strand is being transcribed, the complementary CTG strand while being single-stranded, folds back and forms a hairpin. On the other hand, the non-transcribed CAG strand in orientation I is less able to form stable hairpins. Additionally, in orientation I the CTG strand is not single-stranded and cannot form stable hairpins because it is "occupied" by the RNA polymerase complex. The model further envisages that while TRS is transcribed, it is also replicated. In this case, the CTG hairpin in orientation II will be bypassed by the DNA polymerase complex during lagging-strand synthesis, and this will lead to deletions. Conversely, deletions in orientation I will be found rarely since there is a lower propensity to form secondary structures on the lagging-strand template by the CAG tracts and thus, no bypass synthesis occurs.

DNA Repair

Methyl-directed mismatch repair (MMR)

In all organisms genomic integrity is normally maintained by a variety of DNA repair pathways, including MMR and nucleotide excision repair (NER).[61] Secondary DNA structures formed during DNA synthesis, especially within single-stranded regions containing repetitive tracts, may be hazardous for genome stability if not removed by repair activities. MMR pathway is a fundamental system involved in maintaining genomic integrity because in addition to correcting mismatched base pairs, it also repairs some nonclassical DNA structures such as small hairpins and unpaired regions within DNA. Upon inactivation of MMR increased heterogeneity is observed at simple repetitive DNA (e.g., mono- and dinucleotides) in bacteria, yeast and mammals. The associations of defective MMR and an elevated genetic instability at simple DNA repeats are particularly strong for hereditary nonpolyposis cancer.

The investigations of a role of methyl-directed mismatch repair in TRS instability were an important step in studying the molecular mechanisms leading to the
accumulation of dynamic mutations among triplet repeats. Notably, studies performed
in bacteria and yeast have identified that MMR had contrasting effects on the genetic stability of TRS.[62-64] Although instability of the TRS is linked to human disorders, functional similarities between the MMR in prokaryotes, lower eukaryotes
and humans justified this study. For example, *E.coli* strains with defective MMR
had a reduced occurrence of large deletions (more than 8 repeats) from plasmids
harboring long CTG/CAG. By contrast, mutations in MMR proteins increased the
frequency of small length changes (less than 8 repeats) in shorter CTG/CAG repeats
in *E.coli* and *S. cerevisiae*.

To clarify these apparently conflicting results, Parniewski et al have used a
variety of lengths of CTG/CAG tracts (ranging from 25 to 175 units) to determine
the effects of MMR on repeat tract stability in *E.coli*.[65] They showed that depending
on the length of repeats the functional MMR proteins act to promote large deletions
(usually more than 8 repeats) in CTG/CAG tracts, but significantly prevent length
changes (both, expansions and deletions) of less than 8 repeats. Not only the length
of the TRS influenced the incidence of deletions in CTG/CAG but also the instability
was dependent on the purity of TRS (i.e., presence of interruptions) as well as the
cell growth conditions. One plausible explanation of this distinctive behavior of the
MMR proteins acting on the TRS is the propensity of the triplet repeats to undergo
different structural transitions depending on length of the repeated motif. Since short
TRS are more likely to form slipped structures as opposed to the long ones, which
will rather tend to assemble into stable hairpins therefore, different local Non-B-
DNA structures may trigger particular cellular mechanisms. Considering this and
results from other groups, we propose a model which links structural properties of
the $(CTG)_n$ to the polymerase pausing and bypass synthesis within DNA tracts being
repaired by the MMR (Fig.1.5). Following the DNA slippage of the complementary
strands in double-stranded TRS region, small loops are formed on both strands and
therefore are recognized by functional MMR proteins. The repair process leads to
the excision of large segments of non-methylated strand spanning a region containing
loopouts and to the formation of single-stranded regions on the complementary strand.
Short single-stranded TRS tracts (of less than approx. 100 units) are much less prone
to form stable hairpins than the long ones and resynthesis of the complementary
strand will result in neither deleted nor expanded TRS. In the absence of functional
MMR proteins, the same tract will be subjected to SILC pathway and consequently
small expansions and deletions will gradually accumulate within the repeated motif
after subsequent generations of cells. Therefore, repair of small loops that could
arise on relatively short CTG/CAG tracts would stabilize the TRS and lack of this
repair function will have an opposite effect. Conversely, the same repair pathway
acts differently on long tracts of the TRS. Following the formation of slipped-stranded
structures, recognition of small loops and excision of one DNA strand by the MMR
protein complex will result in long single-stranded stretches in the CTG region which
will self-pair and form stable hairpins. If during resynthesis of a gap DNA

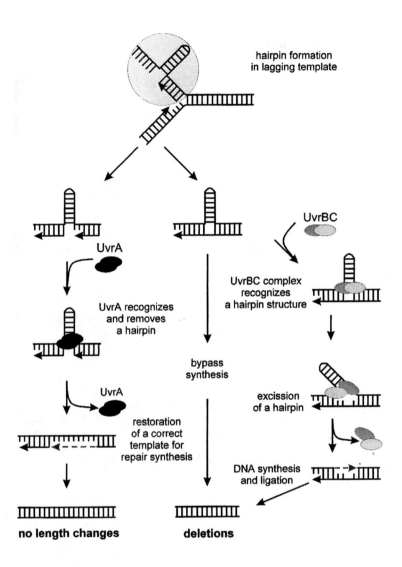

Figure 6. The possible pathways of the NER-generated genetic instability of long transcribed CTG/CAG tracts in orientation II. The CTG hairpin formed on the lagging strand template during TRS replication (shaded circle) may be removed by the UvrA dimer (left panel) and once the correct template for the repair synthesis is restored, the TRS is replicated with no length changes. Bypass synthesis in NER deficient strains will result in large deletions (middle panel). In the absence of the functional UvrA protein the UvrBC complex may specifically recognize and excise the CTG hairpin, which would also lead to large deletions.

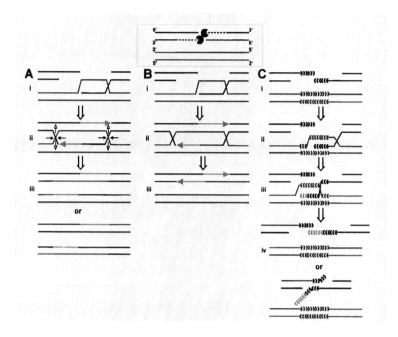

Figure 7. Recombination pathways of Double-strand break repair. (Frame) Double-strand breaks initiate nearly all homologous recombination pathways and are the start point for the 5'→3' exonucleolytic digestion which leaves 3' ends of donor DNA duplex free to invade the template molecule. (A) Szostak et al. model.[75] Invasion leads to the displacement of a template strand (D-loop formation) [i]. Free 3' ends of the donor molecule are the priming sites for the DNA synthesis followed by the formation of „Holliday junctions". Resolution of these structures occurs by cutting each junction in one of two directions (open and closed arrowheads) [ii]. Cutting both junctions in the same orientation results in gene conversion while the opposite cuts yield gene conversion associated with crossover [iii]. (B) Synthesis-Dependent Strand Annealing (SDSA) model. After invasion and DNA synthesis donor strands unwind from the template and reanneal without crossed-over product formation [iii]. (C) Alternative, Bubble Migration SDSA model. Only one 3' ended donor strand invades template molecule [ii]. DNA synthesis occurs within migrating bubble formed by the displaced strand [iii]. After unwinding from the template donor strands containing repeated sequences (multiple arrowheads) may reanneal in out-of-frame order what may lead to expansions or deletions [iv]. Several modifications of SDSA as well as other recombination pathways may be involved in repeated sequences instability (for a review see ref. 74).

Note that in case of the repeated sequences, polymerase slippage and idling may occur during repair synthesis, which may additionally increase the rate of expansions.

polymerase bypasses the hairpin the "repaired" molecule would contain big deletions. However, when the CAG strand serves as a template for repair synthesis (inverse orientation of CTG/CAG tract) the nascent DNA would be able to produce stable hairpins, which would possibly, cause DNA polymerase to stall. Further multiple polymerase slippages, the relocation of newly synthesized repeated DNA fragment and idling synthesis will result in large expansions of the TRS.

Our model presented above explains opposite results concerning the role of the MMR system in generating TRS instabilities in bacteria and yeast, obtained in different laboratories. However, how MMR affects the frequency of expansion events in humans remains unclear. Moreover, long CAG/CTG repeats from the gene associated with Huntington's disease in humans were shown to be less prone to expand in transgenic mice with defective MSH2 protein.[66] Together, these in vivo observations suggest that mutations in MMR enzymes are not required for expansions of TRS in mammals, and the involvement of this repair system in TRS related diseases needs to be more extensively studied.

Nucleotide excision repair (NER)

Nucleotide excision repair is another major cellular defense system in both prokaryotes and eukaryotes. This pathway efficiently recognizes and repairs a vast majority of damages, including bulky DNA adducts and DNA cross-links that cause significant distortion of the helix, as well as less distortive lesions such as methylated bases. Also, the involvement of NER in the repair of DNA loops in vitro has been reported.[67,68] In humans, defects in NER proteins cause at least three hereditary disorders, including xeroderma pigmentosum, Cockayne's syndrome and trichothiodystrophy.[61]

Since unusual DNA structures can form in some TRS in vivo and may therefore invoke destabilization of double-stranded helix, they are also likely to trigger the NER proteins and during the repair process enhance repeat tract instability. Studies in *E.coli* revealed that bacterial NER proteins influence the genetic stability of the TRS in a complex manner.[69] First of all, the stability, as demonstrated by previous investigations was highly dependent on the length of the repeated tract and the orientation of TRS insert relative to the origin of replication. The instability was only observed for long CTG/CAG tracts (175 units) in orientation II, where the CTG strand served as a template for lagging strand synthesis. However, in long-term (multigenerational) growth of the wild type strain and its isogenic *uvrA* or *uvrB* mutants, the rate of deletions in strain lacking functional UvrA protein was significantly higher as compared to strain that lacked only UvrB. In *E.coli* UvrA is required for damage recognition. The affinity of the UvrA protein to single-stranded DNA, specifically to bubbles and loops may be responsible for the recognition and binding to the CTG hairpins in their single-stranded loop region.[67,70] Binding of the UvrA to unusual conformations may destabilize such structures allowing the correct copying of the entire repeat. Others have demonstrated that absence of the single-stranded-DNA-binding protein (SSB) in vivo similarly led to an increased frequency of large deletions within the triplet repeats.[71] Very high stability of long CTG/CAG tracts in strains lacking functional UvrB suggests that this protein may be involved in processing of unusual structures within repeats and allows deletions to occur. In some in vitro studies specific recognition and excision of bubbles within double-stranded DNA by the UvrBC endonucleolytic complex was demonstrated.[72] An alternative scenario is that in the absence of the UvrA protein, the CTG hairpin may be also a substrate for the cellular endonucleolytic activities. Such nicked DNA may be degraded in vivo which would similarly lead to deletions. The possible pathways of the CTG hairpins processing by NER are presented on Figure 1.6.

Interestingly, the genetic differences in the stability of long CTG/CAG tracts between *uvrA* and *uvrB* mutants were apparent only if the TRS were transcribed. Transcription through the TRS may additionally stabilize CTG hairpins by introducing negative supercoils behind RNA polymerase complex. It is important to note that the NER pathway is well suited to repair transcribed strands. Any kind of RNA polymerase pausing triggers transcription-repair coupling factor (TRCF). This protein attracts NER components to the transcribed region providing prompt removal of DNA lesions. One might assume that formation of the hairpin structures on the template strand as well as on the nascent RNA may lead to RNA polymerase stalling. Napierala et al demonstrated that CUG repeats form extremely stable, length-dependent, self-complementary structures. This strongly supports the hypothesis that structural aberrations within the TRS are causative for their genetic instability.

Recombination

Recombination is a major source of the genetic instability of all organisms. This process allows the cell to change the order of its genes, to move the sequence from one place to another (translocations), change the orientation (inversions), multiply (duplications) or remove (deletions) from the genome. However, one must remember that it serves also very often to repair damaged DNA.

Several pathways of recombination are known of which the most frequent one is the homologous recombination dependent on RecA protein in case of bacteria or its eukaryotic equivalents (i.e., Rad51p family in yeast). This process occurs usually between very similar or identical DNA tracts located on two different DNA molecules and positioned in the same regions with respect to the entire molecule (allelic recombination) and basically serves to keep the genome stable. Sometimes however, it may involve the sequences dispersed among the same chromosome for example direct and inverted repeats or the sequences located on nonhomologous replicons (ectopic or homeologous recombination). Interactions between such sequences lead to the gross genome rearrangements mentioned above.

The stability of the repeating sequences such as micro- and minisatellites has been investigated for several years and recently recombination has been shown as the second (along with the replication) major mechanism responsible for the contractions as well as the expansions of such tracts. This progress has been made particularly due to the improvement of the in vivo genetic assays allowing precise investigations of several pathways of recombination in eukaryotic cells, especially in yeast.

Among different recombination pathways, gene conversion, i.e., nonreciprocal transfer of the genetic information from one DNA duplex to another leading to non-Mendelian segregation is suggested to be the major source of the repeating sequences instability. This process is sometimes associated with cross-over events and the proportion of the gene conversions that are accompanied by the crossing over seems to be much higher during meiosis than mitosis.[73,74] Different groups have explained mechanisms of gene conversion occurring without as well as with gene crossing-over, although the common feature of all models was the initiation of the process by the double strand break (DSB) within one duplex of DNA. A complex model has been proposed by Szostak et al[75] (Fig.1.7, panel A). This model assumes that DSB formation followed by the exonucleolytic digestion of the 5' ends leads to the formation of large gaps with 3' overhangs (up to 1kb or even more) which can invade a homologous template in order to repair broken DNA.[76] Both 3' ends serve as the priming site for DNA polymerase, which elongates them using undamaged duplex as the template. After formation of the Holliday junctions, these four-stranded, branched structures may migrate in both directions spanning bigger regions and ultimately may be resolved by cutting in one of two orientations. If the noncrossover strands are cut in one Holliday junction and crossed strands are cut in another, this gives rise to the gene conversion associated with the crossing over. Alternatively, if both junctions are cut in the same orientation, resulting gene conversion will not be associated with crossover. Such a concept assumes that the ratio between both types of events should be equal. Although several experimental data support this model, some work has revealed that this ratio is strongly biased towards the noncrossover products, especially during mitosis.[77-79] One explanation could be that there is some preference during the resolution of the Holliday junctions which leads to the same way of cutting in both structures (i.e., the resolution requires isomerization which allows only crossed strands to be cleaved).[80,81] Other alternatives assuming that there is no need for Holliday junction formation in order to perform gene conversion have been proposed based on the analysis of the recombination products in such different organisms as *E.coli*, yeast, *Ustilago*, *Drosophila* as well as in humans.[78,79,82-86] This alternative model of DSB repair called Synthesis-Dependent Strand Annealing (SDSA) assumes that after strand invasion both 3' ends of the donor are extended by DNA polymerase while the donor DNA remains unchanged.[74,87] Thus, in contrast to the commonly known semiconservative character of the DNA replication, here the strand synthesis becomes a conservative process as it occurs on two strands within one duplex. The newly synthesized strands unwind from their templates and reanneal back within the broken duplex (Fig.1.7, panel B). Both unique as well as the repeated sequences may recombine via this pathway, although in the case when the repeated sequences are involved, expansions and deletions may occur. This is due to the fact that after elongation newly synthesized strands contain two or more repeats which during reannealing may pair in out of frame order (Fig.1.7, panel C). One must mention that SDSA models, although they explain the bias towards noncrossover events, do not exclude the possibility of crossover occurrence. If both strands of the recipient molecule are used as a synthesis start point, then it may lead to the formation of the two Holliday junctions so the gene conversion may be followed by the crossover as in the classical view of Szostak's model.

For several years the mechanism of general homologous recombination has been considered to play an important role also in the instability of the trinucleotide repeats. Although the crossover mechanism seemed to be rather unlikely as no exchange of the flanking sequences has been observed, other recombination events like gene conversion or unequal crossover were suggested to be responsible for the triplet repeats size alterations observed in humans, mainly in myotonic dystrophy and fragile X patients.[88-92] Triplet repeats, although belonging to the category of microsatellites, feature some distinct characteristics which make them specific substrates for the recombination machinery. It has been established that double strand breaks may occur very frequently within these sequences. One reason for this could be the fact that the replication fork moving across long TRS may frequently pause leading to the formation of the unfinished Okazaki fragments.[40,41] Such regions may induce the formation of the double strand breaks. The process of DNA polymerase stalling facilitates also the formation of the secondary structures on such incompletely synthesized Okazaki fragments. On the other hand, such structures may also form on the template strand, as it remains single-stranded. Besides the important role of the replication-based instability, which happens during formation of alternative secondary conformations (polymerase slippage), these structures also may be recognized by the specific cellular endonucleases resulting in DSB formation and subsequent increased recombination rates.[93,94]

Many experimental assays have been developed recently in order to analyze specifically the destabilizing effect of the recombination machinery on TRS. In bacteria, a system involving two-plasmid model has been used where two otherwise nonhomologous vectors both carrying TRS of different length were introduced into the recA$^+$ background.[95,96] Analysis of the recombination products using restriction mapping and sequencing revealed that CTG/CAG tracts were better substrate for gene conversion mechanisms than CGG/CCG tracts. Only long CTG/CAG tracts (more than 30 repeats) were recombinogenic and yielded with the multiple-fold-expanded products. Moreover, the recombination-induced expansions were much more frequent than deletions (approximately 100:1 ratio)—in strong opposition to the results obtained when the TRS instability was induced by the replication mechanisms (1:100 ratio). The observation obtained with this model has not yet been confirmed though, as the attempts of another group to get CTG/CAG expansions using the similar system showed, there was a replication—but not recombination—dependent character of the expansions.[97]

Several systems have been also established in order to observe TRS instability in yeast. Fungi historically served as the model organism in the recombination study as they allow for easy analysis of the products formed in mitosis as well as in meiosis. Therefore they are ideally suited for investigations of TRS instability as it is now thought that this process takes place in humans during meiosis as well as in early postzygotic stages of mitotic cell growth.[98] It has been shown that most mitotic and basically all meiotic recombination processes in yeast are induced by double-strand breaks although the proteins as well as the mechanism of recombination occurring during these events are different.[74] Indeed, the instability of TRS was observed as the recombination outcome of both types of divisions, although CTG/CAG tracts were much more prone to give deletions as well as expansions during meiosis, while mostly deletions were observed after mitosis.[99,100] This process was dependent on the activity of the topoisomerase II-like transesterase—Spo11, which induces the formation of the DSB within triplet repeats during meiosis.[101] Only long tracts were recombinogenic while the shorter ones (10 repeats) were much more stable.[99, 102] Analysis of the mitotic recombination events revealed that the bias towards expansions or deletions of the CAG/CTG repeats was dependent on the length of such regions. Shorter tracts (39 repeats) yielded contractions and the longer ones (98 repeats) showed both deletions as well as expansions.[103] On the other hand, in the case of CCG/CGG repeats, no difference in stability was observed between mitotic and meiotic events regardless on the length of the repeated tracts.[104]

The study of the recombination-induced TRS instabilities are also currently ongoing using mammalian models including human cells. Several yeast protein homologues involved in the general homologous recombination between TRS as well as other sequences have been identified in higher eukaryotes. Some of them seem to play an even more essential role than in yeast since their absence may give very severe phenotypes in vertebrates. Among them, homologues of Rad51p (which deletion is lethal for vertebrates but not for yeast) and its related proteins Rad55p and Rad 57p, as well as Dmc1, Rad54p, Rad52p, Mre11, Rad50, Xrs11 have been described. Also the homologues of the yeast proteins engaged specifically in the meiotic recombination like Spo11, Msh4p, Msh5p have been identified (for a review see refs. 74, 105). Analysis of the mutants lacking such proteins indicates that in the higher eukaryotes they play a similar role (although in some cases a slightly distinct role) and thus one may predict their involvement in the instability of the triplet-repeat sequences in humans. Nevertheless, more specific data on the recombination-induced TRS expansions based on the analysis of vertebrate models, especially human cells, is highly desired.

It is also worth mentioning that instability of the repeated sequences (possibly including TRS) may result from the activity of recombination pathways not described in this section. For example, some experiments have demonstrated that the recombination between direct repeats located in close proximity can occur efficiently in a RecA-independent manner in *E.coli* cells. The frequency of this process was dependent on the length of the direct repeats as well as the distance between them. Moreover, these factors had the impact on the outcome of the process as the short neighboring repeats yielded monomeric products, whereas the increase of length as well as the distance between the repeats gave rise to the two different kinds of dimeric products. The model in which the misalignment of the direct repeats during DNA replication forms highly recombinogenic substrate, which can be then processed by different RecA-independent pathways, may explain this observation.[106]

Finally, the involvement of the illegitimate recombination induced by side-products of the activity of some enzymes (i.e., topoisomerases) should be taken into consideration.

SUMMARY

To date several neurodegenerative disorders including myotonic dystrophy, Huntington's disease, Kennedy's disease, fragile X syndrome, spinocerebellar ataxias or Friedreich's ataxia have been linked to the expanding trinucleotide sequences. Although phenotypic features vary among these debilitating diseases, the structural abnormalities of the triplet repeat containing DNA sequences is the primary cause for all of these disorders. Expansions of the CAG repeat within coding regions of miscellaneous genes result in the synthesis of aberrant proteins containing enormously long polyglutamine stretches. Such proteins acquire toxic functions and/or may direct cells into the apoptotic cycle. On the other hand, massive expansions of various triplet repeats (i.e., CTG/CAG, CGG/CCG/, GAA/TTC) inside the noncoding regions lead to the silencing of transcription and therefore affect expression of the adjacent genes. The repetitive character of TRS allows stretches of such tracts to form slipped-stranded structures, self-complementary hairpins, triplexes or more complex configurations called "sticky DNA", which are not equally processed by some cellular mechanisms, as compared to random DNA.

It is likely that the instability of the short TRS (below the threshold level) occurs due to the SILC pathway, which is driven by the DNA slippage. Accumulation of the short expansions leads to the disease premutation state where the MLC pathway becomes predominant. Independent of which mechanism is involved in the MLC pathway (replication, transcription, repair or recombination) the process of complementary strand synthesis is crucial for the TRS instability. Generally, dependent on the location of the tract which has higher potential to form secondary DNA structure, further processing of such tract may result in expansions (secondary structure formed at the newly synthesized strand) or deletions (structure present on the template strand).

Analyses of molecular mechanisms of the TRS genetic instability using bacteria, yeast, cell lines and transgenic animals as models allowed the scientists to better understand the role of some major cellular processes in the development of neurodegenerative disorders in humans. However, it is necessary to remember that most of these investigations were focused on the involvement of each particular process separately. Much less of this work though was dedicated to the search for the interactions between such cellular systems that in effect could result in different rate of TRS expansions. Thus, more intensive studies are necessary in order to fully understand the phenomenon of the dynamic mutations leading to the human hereditary neurodegenerative diseases.

ACKNOWLEDGMENTS

Authors were partially supported by the State Committee for Scientific Research (KBN grant 6P04A 01617).

REFERENCES

1. Charlesworth B, Sniegowski P, Stephan W. The evolutionary dynamics of repetitive DNA in eukaryotes. Nature 1994; 371:215-220.
2. Moxon ER, Wills C. DNA microsatellites: Agents of evolution? Sci Am 1999; January:94-99.
3. International Human Genome Sequencing Consortium. Initial sequencing and analysis of the human genome. Nature 2001; 409:860-921.
4. Venter JC, Adams MD, Myers EW et al. The sequence of the human genome. Science 2001; 291(5507):1304-1351.
5. Wells RD, Warren ST. Genetic Instabilities and Hereditary Neurological Diseases. San Diego: Academic Press, 1998.
6. Oostra BA. Trinucleotide Diseases and Instability. Heidelberg: Springer-Verlag, 1998.
7. Cummings CJ, Zoghbi HY. Fourteen and counting: Unraveling trinucleotide repeat diseases. Hum Mol Genet 2000; 9:909-916.
8. Richards RI, Sutherland GR. Dynamic mutations: A new class of mutations causing human disease. Cell 1992; 70:709-712.
9. Sutherland GR, Richards RI. Simple tandem DNA repeats and human genetic disease. Proc Natl Acad Sci USA 1995; 92:3636-3641.
10. McInnis MG. Anticipation: An old idea in new genes. Am J Hum Genet 1996; 59:973-979.
11. McMurray CT. DNA secondary structure: A common and causative factor for expansion in human disease. Proc Natl Acad Sci USA 1999; 96:1823-1825.
12. Modrich P, Lahue R. Mismatch repair in replication fidelity, genetic recombination and cancer biology. Annu Rev Biochem 1996; 65:101-133.
13. Prolla TA. DNA mismatch repair and cancer. Curr Opin Cell Biol 1998; 10:311-316.
14. Umar A, Kunkel TA. DNA-replication fidelity, mismatch repair and genome instability in cancer cells. Eur J Biochem 1996; 238:297-307.
15. Fishel R, Kolodner RD. Identification of mismatch repair genes and their role in the development of cancer. Curr Opin Genet & Dev 1995; 5:382-395.
16. Lengauer CK, Kinzler W, Vogelstein B. Genetic instability in colorectal cancers. Nature 1997; 386:623–627.
17. Schlötterer C. Evolutionary of microsatellite DNA. Chromosoma 2000; 109:365-371.
18. Levinson G, Gutman GA. Slipped-strand mispairing: A major mechanism for DNA sequence evolution. Mol Biol & Evol 1987; 4:203-221.
19. Sinden RR, Pearson CE, Potaman VN et al. DNA: Structure and function. Adv Genome Biol 1998; 5A:1-141.
20. Jaworski A, Hsieh WT, Blaho JA et al. Left-handed DNA in vivo. Science1987; 238(4828):773-777.
21. Panayotatos N, Fontaine A. A native cruciform DNA structure probed in bacteria by recombinant T7 endonuclease. J Biol Chem 1987; 262(23):11364-11368.
22. Pearson CE, Sinden RR. Alternative structures in duplex DNA formed within the trinucleotide repeats of the myotonic dystrophy and fragile X loci. Biochemistry1996; 35(15):5041-5053.
23. Parniewski P, Galazka G, Wilk A et al. Complex structural behavior of oligopurine-oligopyrimidine sequence cloned within the supercoiled plasmid. Nucl Acids Res 1989; 17(2):617-629.

24. Parniewski P, Kwinkowski M, Wilk A et al. Dam methyltransferase sites located within the loop region of the oligopurine-oligopyrimidine sequences capable of forming H-DNA are undermethylated in vivo. Nucl Acids Res 1990; 18 (3):605-611.

25. Kohwi Y, Malkhosyan SR, Kohwi-Shigematsu T. Intramolecular dG.dG.dC triplex detected in *Escherichia coli* cells. J Mol Biol 1992; 223(4):817-822.

26. Sen D, Gilbert W. Formation of parallel four-stranded complexes by guanine-rich motifs in DNA and its implications for meiosis. Nature1988; 334(6180):364-366,

27. Mitas M. Trinucleotide repeats associated with human disease. Nucl Acids Res 1997; 25:2245-2254.

28. Usdin K, Woodford KJ. CGG repeats associated with DNA instability and chromosome fragility form structures that block DNA synthesis in vitro. Nucl Acids Res 1995; 23:4202-4209.

29. Gacy AM, Goellner GM, Spiro C et al. GAA instability in Friedreich's ataxia shares a common, DNA-directed and intraallelic mechanism with other trinucleotide diseases. Mol Cell 1998; 1:583-593.

30. Smith GK, Jie J, Fox GE et al. DNA CTG triplet repeats involved in dynamic mutations of neurologically related gene sequences for stable duplexes. Nucl Acids Res 1995; 23:4303-4311.

31. Mariappan SV, Catasti P, Chen X et al. Solution structures of the individual single strands of the fragile X DNA triplets (GCC)n/(GGC)n. Nucl Acids Res 1996; 24(4):784-92.

32. Bacolla A, Gellibolian R, Shimizu M et al. Flexible DNA: Genetically unstable CTG◊CAG and CGG◊CCG from human hereditary neuromuscular disease genes. J Biol Chem 1997; 272:16783-16792.

33. Gellibolian R, Bacolla R, Wells RD. Triplet repeat instability and DNA topology: An expansion model based on statistical mechanics. J Biol Chem 1997; 272:16793-16797.

34. Wang Y-H, Amirhaeri S, Kang S et al. Preferential nucleosome assembly at DNA triplet repeats from the myotonic dystrophy gene. Science 1994; 265:669-671.

35. Wang Y-H, Griffith JD. Expanded CTG triplet blocks from the myotonic dystrophy gene create the strongest known natural nucleosome positioning elements. Genomics 1995; 25:570-573.

36. Godde JS, Wolffe AP. Nucleosome assembly on CTG triplet repeats. J Biol Chem 1996; 271:15222-15229.

37. Wang Y-H, Gellibolian R, Shimizu M et al. Long CCG triplet repeat blocks exclude nucleosomes: A possible mechanism for the nature of fragile sites in chromosomes. J Mol Biol 1996; 263:511-516.

38. Godde JS, Kass SU, Hirst MC et al. Nucleosome assembly on methylated CGG triplet repeats in the Fragile X mental retardation gene 1 promoter. J Biol Chem 1996; 271:24325-24328.

39. Otten AD, Tapscott SJ. Triplet repeat expansion in myotonic dystrophy alters the adjacent chromatin structure. Proc Natl Acad Sci USA 1995; 92:5465-5469.

40. Samadashwily G, Raca G, Mirkin SM. Trinucleotide repeats affect DNA replication in vivo. Nat Genet 1997; 17:298-304.

41. Kang SM, Ohshima K, Shimizu M et al. Pausing of DNA synthesis in vitro at specific loci in CTG and CGG triplet repeats from human hereditary disease genes. J Biol Chem 1995; 270:27014-27021.

42. Pluciennik A, Iyer RR, Parniewski P et al. Tandem duplications: A novel type of triplet repeat instability. J Biol Chem 2000; 275(37):28386-28397.

43. Gururajan R, Lahti JM, Grenet J et al. Duplication of a genomic region containing Cdc2L1-2 and MMP21-22 genes on human chromosome 1p36.3 and their linkage to D1Z2. Genet Res 1998; 8:929-939.

44. Tautz D, Schlötterer C. Simple sequences. Curr Opin Genet & Dev 1994; 4:832-837.

45. Lovett ST, Feschenko VV. Stabilization of diverged tandem repeats by mismatch repair: Evidence for deletion formation via a misaligned replication intermediate. Proc Natl Acad Sci USA 1996; 93:7120-7124.

46. Wells RD, Parniewski P, Pluciennik A et al. Small slipped register genetic instabilities in *Escherichia coli* in triplet repeat sequences associated with hereditary neurological diseases. J Biol Chem 1998; 273:19532-19541.

47. Sakamoto N, Chastain PD, Parniewski P et al. Sticky DNA: Self-association properties of long GAA·TTC repeats in R·R·Y triplex structures from Friedreich's ataxia. Mol Cell 1999; 3:465-475.

48. Bidichandani SI, Ashizawa T, Patel PI. The GAA triplet-repeat expansion in Friedreich ataxia interferes with transcription and may be associated with an unusual DNA structure. Am J Hum Genet 1998; 62:111-121.

49. Ohshima K, Montermini L, Wells RD et al. Inhibitory effects of expanded GAA/TTC triplet repeats from intron I of the Friedreich's ataxia gene on transcription and replication in vivo. J Biol Chem 1998; 273:14588-14595.

50. Kang S, Jaworski A, Ohshima K et al. Expansion and deletion of CTG repeats from human disease genes are determined by the direction of replication in *E.coli*. Nature Genet 1995; 10:213-218.

51. Ohshima K, Kang S, Wells RD. CTG triplet repeats from human hereditary diseases are dominant genetic expansion products in *Escherichia coli*. J Biol Chem 1996; 271:1853-1856.

52. Kang S, Ohshima K, Jaworski A et al. CTG triplet repeats from the myotonic dystrophy gene are expanded in *Escherichia coli* distal to the replication origin as a single large event. J Mol Biol 1996; 258:543-547.

53. Iyer RR, Wells RD. Expansion and deletion of triplet repeat sequences in *Escherichia coli* occur on the leading strand of DNA replication. J Biol Chem 1999; 274:3865-3877.

54. Sarkar PS, Chang H-C, Boudi FB et al. CTG repeats show bimodal amplification in *E.coli*. Cell 1998; 95:531-540.

55. Freudenreich CH, Stavenhagen JB, Zakian VA. Stability of a CTG/CAG trinucleotide repeat in yeast is dependent on its orientation in the genome. Mol Cell Biol 1997; 17:2090-2098.

56. Schweitzer JK, Livingston DM. Expansions of CAG repeat tracts are frequent in a yeast mutant defective in Okazaki fragment maturation. Hum Mol Genet 1998; 7:69-74.

57. Liu LF, Wang JC. Supercoiling of the DNA template during transcription. Proc Natl Acad Sci USA 1987; 84:7024-7027.

58. Bowater RP, Jaworski A, Larson JE et al. Transcription increases the deletion frequency of long CTG·CAG triplet repeats from plasmids in *Escherichia coli*. Nucl Acids Res 1997; 25:2861-2868.

59. Parniewski P, Bacolla A, Jaworski A et al. Nucleotide excision repair affects the stability of long transcribed (CTG·CAG) tracts in an orientation dependent manner in *Escherichia coli*. Nucl Acids Res 1999; 27:616-623.

60. Wierdl M, Greene CN, Datta A Destabilization of simple repetitive DNA sequences by transcription in yeast. Genetics 1996; 143:713-721.

61. Friedberg EC, Walker GC, Siede W. DNA Repair and Mutagenesis. Washington DC: American Society for Microbiology, 1995.

62. Jaworski A, Rosche WA, Gellibolian R et al. Mismatch repair in *Escherichia coli* enhances instability of (CTG)$_n$ triplet repeats from human hereditary diseases. Proc Natl Acad Sci USA 1995; 92:11019-11023.

63. Schumacher S, Fuchs RPP, Bichara M. Expansion of CTG repeats from human disease genes is dependent upon replication mechanisms in *Escherichia coli*: The effect of long patch mismatch repair revisited. J Mol Biol 1998; 279:1101-1110.

64. Schweitzer JK, Livingston DM. Destabilization of CAG trinucleotide repeat tracts by mismatch repair mutations in yeast. Hum Mol Gen 1997; 6:349-355.

65. Parniewski P, Jaworski A, Wells RD et al. Length of CTG.CAG repeats determines the influence of mismatch repair on genetic instability. J Mol Biol 2000; 299(4):865-874.

66. Manley K, Shirley TL, Flaherty L et al. MSH2 deficiency prevents in vivo somatic instability of the CAG repeat in Huntington disease transgenic mice. Nature Genetics 1999; 23(4):471-473.

67. Ahn B, Grossman L. The binding of UvrAB proteins to bubble and loop regions in duplex DNA. J Biol Chem 1996; 271(35):21462-21470.

68. Kirkpatrick DT, Petes TD. Repair of DNA loops involves DNA-mismatch and nucleotide-excision repair proteins. Nature1997; 387(6636):929-931.

69. Parniewski P, Bacolla A, Jaworski A et al. Nucleotide excision repair affects the stability of long transcribed (CTG*CAG) tracts in an orientation-dependent manner in *Escherichia coli*. Nucl Acids Res 1999; 27(2):616-623.

70. Seeberg E, Steinum AL. Purification and properties of the UvrA protein from *Escherichia coli*. Proc Natl Acad of Sci USA 1982; 79(4):988-992.

71. Rosche WA, Jaworski A, Kang S et al. Single-stranded DNA-binding protein enhances the stability of CTG triplet repeats in *Escherichia coli*. J Bacteriol 1996; 178:5042-5044.

72. Zou Y, Walker R, Bassett H et al. Formation of DNA repair intermediates and incision by the ATP-dependent UvrB-UvrC endonuclease. J Biol Chem 1997; 272(8):4820-4827.

73. Fogel S, Mortimer RK, Lusnak K. Mechanisms of gene conversion, or wandering on a foreign strand. In: Strathern JN, Jones EW, Broach JR, eds. The Molecular Basis of the Yeast *Saccharomyces cerevisisae*: Life and Inheritance. Cold Spring Harbor: Cold Spring Harbor Laboratory, 1981:289-339.

74. Paques F, Haber JE. Multiple pathways of recombination induced by double-strand breaks in *Saccharomyces cerevisisae*. Microbiol Mol Biol Rev 1999; 63(2):349-404.

75. Szostak JW, Orr-Weaver TL, Rothstein RJ et al. The double-strand break repair model for recombination. Cell 1983; 33:25-35.

76. Sun H, Treco D, Szostak JW. Extensive 3' overhanging single-stranded DNA associated with the meiosis-specific double-strand breaks at the ARG4 recombination initiation site. Cell 1991; 64:1155-1161.

77. Orr-Weaver TL, Szostak JW. Yeast recombination: the association between double-strand gap repair and crossing-over. Proc Natl Acad Sci USA 1983; 80:4417-4421.

78. Klein HL. Lack of association between intrachromosomal gene conversion and reciprocal gene exchange. Nature 1984; 310:748-753.

79. Plessis A, Dujon B. Multiple tandem integrations of transforming DNA sequences in yeast chromosomes suggest a mechanism for integrative transformation by homologous recombination. Gene 1993; 134:41-50.

80. Meselson MM, Radding CM. A general model for genetic recombination. Proc Natl Acad Sci USA 1975; 72:358-361.

81. Bennett RJ, West SC. RuvC protein resolves Holliday junctions via cleavage of the continous (noncrossover) strands. Proc Natl Acad Sci USA 1995; 92:5635-5639.

82. Kreuzer KN, Saunders M, Weislo LJ et al. Recombination-dependent DNA replication stimulated by double-strand breaks in bacteriophage T4. J Bacteriol 1995; 177:6844-6853.

83. Fergusson DO, Holloman WK. Recombinational repair of gaps in DNA is asymmetric in *Ustilago maydis* and can be explained by a migrating D-loop model. Proc Natl Acad Sci USA 1996; 93:5419-5424.

84. Engels WR, Johnson-Schlitz DM, Eggleston WB et al. High-frequency P-element loss in *Drosophila* is homolog dependent. Cell 1990; 62:515-525.

85. Gloor GB, Nassif NA, Johnson-Schlitz DM et al. Targeted gene replacement in *Drosophila* via Pelement-induced gap-repair. Science 1991; 253:533-540.

86. effreys AJ, Tamaki K, MacLeod A et al. Complex gene conversion events in germline mutation at human minisatellites. Nat Genet 1994; 6:136-145.

87. Nassif N, Penney J, Pal S et al. Efficient copying of nonhomologous sequences from ectopic sites via P-element-induced gap repair. Mol Cel Biol 1994; 14:1613-1625.

88. Sinden RR, Wells RD. DNA structure, mutations, and human genetic disease. Curr Opin Biotechnol 1992; 3:612-622.

89. O'Hoy KL, Tsilfidis C, Mahadevan MS et al. Reduction in size of the myotonic dystrophy trinucleotide repeat mutation during transmission. Science 1993; 259:809-810.

90. Jansen G, Willems P, Coerwinkel M et al. Gonosomal mosaicism in myotonic dystrophy patients: Involvement of mitotic events in (CTG)n repeat variation and selection against extreme expansion in sperm. Am Soc Hum Genet 1994; 54:575-585.

91. Van den Ouweland AMW, Deelen W II, Kunst CB et al. Loss of mutation at the FMR1 locus through multiple exchanges between maternal X chromosomes. Hum Mol Genet 1994; 3:1823-1827.

92. Brown WT, Houck GE Jr, Ding X et al. Reverse mutations in the fragile X syndrome. Am J Med. Genet 1996; 64:287-292.

93. Gordenin DA, Kunkel TA, Resnick MA. Repeat expansion—All in a flap? Nat Genet 1997; 16:116-118.

94. Nag DK, Kurst A. A 140-bp-long palindromic sequence induces double-strand breaks during meiosis in the yeast *Saccharomyces cerevisiae*. Genetics 1997; 146:835-847.

95. Jakupciak JP, Wells RD. Genetic instabilities in (CTG•CAG) repeats occur by recombination. J Biol Chem 1999; 274(33):23468-23479.

96. Jakupciak JP, Wells RD. Gene conversion (recombination) mediates expansions of CTG•CAG repeats. J Biol Chem 2000; 275(51):40003-40013.

97. Sopher BL, Myrick SB, Hong JY et al. In vivo expansion of trinucleotide repetas yields plasmid and YAC constructs for targeting and transgenesis. Gene 2000; 261:383-390.

98. McMurray CT. Mechanisms of DNA expansion. Chromosoma 1995; 104(1):2-13.

99. Jankowski C, Nasar F, Nag DK. Meiotic instability of CAG repeat tracts occurs by double-strand break repair in yeast. Proc Natl Acad Sci USA 2000; 97(5):2134-2139.

100. Cohen H, Sears DD, Zenwirth D et al. Increased instability of human CTG repeat tracts on yeast artificial chromosomes during gametogenesis. Mol Cell Biol 1999; 19:4153-4158.

101. Keeney S, Giroux CN, Kleckner N. Meiosis-specific DNA double-strand breaks are catalyzed by Spo11, a member of a widely conserved protein family. Cell 1997; 88:375-384.

102. Moore H, Greenwell PW, Liu C-P et al. Triplet repeats form secondary structures that escape DNA repair in yeast. Proc Natl Acad Sci USA 1999; 96:1504-1509.

103. Paques F, Leung W-Y, Haber JE. Expansions and contractions in a tandem repeat induced by double strand break repair. Mol Cell Biol 1998; 18(4):2045-54.

104. White PJ, Borts RH, Hirst MC. Stability of human fragile X (CGG)n triplet repeat array in *Saccharomyces cerevisiae* deficient in aspects of DNA metabolism. Mol Cell Biol 1999; 19(8):5675-5684.

105. Haber JE. Recombination: A frank view of exchanges and vice versa. Curr Opin Cell Biol 2000; 12:286-292.

106. Bi X, Liu LF. A replication model for DNA recombination between direct repeats. J Mol Biol 1996; 256(5):849-858.

MYOTONIC DYSTROPHY: DISCUSSION
OF MOLECULAR BASIS

Lubov T. Timchenko,[1] Steve J. Tapscott,[2] Thomas A. Cooper[3]
and Darren G. Monckton[4]

Myotonic dystrophy 1 (DM1) is a dominant, neuromuscular disease which represents the most common form of muscular dystrophy with a frequency of 1 in 8,000. Today, there is no cure for this disease. Clinical manifestations vary from the almost asymptomatic condition to the deadly form of disease associated with increased disease severity in generations with reduction of age of onset.

Identification of the gene responsible for the disease and discovery of a CTG trinucleotide expansion as the mutation for DM1 explained many aspects of the clinical features of DM1. However, the development of treatment requires elucidation of the molecular mechanisms of DM1 pathogenesis, explaining how the increase of the length of CTG repeats induces disease. The solution of this problem is complicated because the CTG expansion is located in the 3′ untranslated region rather than in the coding region of the mutant gene. Because of this unusual feature of the DM1 mutation it took almost a decade to reveal the most significant features of the molecular pathogenesis of DM1. This review focuses on the latest data related to the molecular basis of instability of CTG repeats and the complex molecular pathogeneses mediated by unstable CTG/CUG repeats in DM1 patients.

DM1 MUTATION IS AN EXPANSION OF CTG TRINUCLEOTIDE REPEATS

DM1 is an autosomal, dominant, neuromuscular disease characterized by involvement of multiple systems.[1] There are two forms of disease, adult and congenital. The adult form is characterized by myotonia, muscle weakness and wasting,

[1]Section of Cardiovascular Sciences, Departments of Medicine, and Molecular Physiology and Biophysics and [3]Departments of Pathology and Molecular and Cell Biology, Baylor College of Medicine, Houston, TX 77030 U.S.A.; [2]Division of Human Biology, Fred Hutchinson Research Center, Seattle, WA 98109 U.S.A.; and [4]Institute of Biomedical and Life Sciences, The University of Glasgow, Glasgow, G12 8QQ U.K.

27

cataracts, cardiac abnormalities, testicular atrophy, and insulin resistance. The most severe form of DM1, congenital DM1, is associated with hypotonia, mental retardation, and delayed muscle maturation.

The CTG Expansion is an Unstable Mutation within the 3' UTR of the DMPK Gene

The CTG triplet repeat expansion in the normal populations consists of 5-37 units, however, in patients with DM1 the length of CTG expansion is significantly increased up to many thousands of repeats. The number of CTG repeats within the DMPK gene positively correlates with the severity of the symptoms and negatively correlates with age, a phenomenon known as "anticipation".

Once into the expanded disease-associated range, the DM1 CTG repeat becomes highly unstable. The germline mutation rate is essentially 100% and highly biased toward further increases in repeat length. Since the length of the repeat is positively correlated with disease severity and inversely correlated with the age of onset of symptoms, these increases account for the high levels of clinical anticipation observed in DM1 families.[2] The length changes observed are usually large such that the repeat rapidly expands from the range associated with the late onset form of the disease (60 to 100 repeats), through the adult onset form of the disease (200 to 500 repeats), to the congenital form of the disease (700 to 4,000 repeats), often in as few as three generations. Small expansions in humans (50 to 80 repeats) are particularly biased toward large expansions in the male germline, accounting for the excess of transmitting grandfathers in DM1 pedigrees.[3] In contrast, larger expansions in the range 200 to 500 repeats, are most likely to expand further when transmitted by a female, explaining the almost complete association between maternal transmission and congenital DM1.[4] Instability in the soma is also extensive and follows reproducible dynamics. Multiple small mutations biased toward expansions occur throughout life resulting in a gradual increase in the level of variation and the average repeat length within a tissue.[5-7] The pattern of variation between different somatic tissues appears to be conserved, but the rate at which variation accumulates is tissue specific. Most interestingly, the repeat length observed in muscle is always much larger than that observed in blood DNA.[5,8-10] This is intriguing considering the tissue specificity of DM1 symptoms and the post-mitotic nature of muscle in the adult. It seems reasonable to assume therefore that age-dependent, tissue-specific, expansion-biased somatic mosaicism contributes toward the progressive nature and tissue specificity of the symptoms. Although analysis of DM1 patient samples has yielded a number of significant insights into the dynamics of the expanded repeats, the utility of such an approach is limited by the availability of appropriate samples and the confounding effects of allele length, age and genetic background.

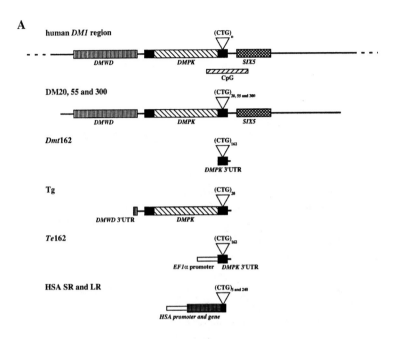

Figure 1A. The human DM1 genomic region and mouse transgenes. Shown are the human *DM1* region including the upstream *DMWD* gene, *DMPK* and the downstream *SIX5* gene, the position of the CTG repeat and the repeat containing CpG island. Also shown are the five transgene constructs used to generate transgenic mice: DM20, 55 and 300 are large cosmid constructs containing all three flanking genes (16, 23, 26); *Dmt*162 comprises only the 3' UTR of *DMPK* (14,27); Tg contains the *DMPK* gene and leads to over expression of DMPK protein (41; *Te*162 drives expression of the *DMPK* 3'UTR from the elongation factor 1 alpha (EF1α) promoter (60) and the HSA SR (short repeat) and LR (long repeat) transgenes drive expression of a CTG tract in the 3'UTR of a human skeletal muscle alpha actin gene (HSA) from the HSA promoter (61).

MOUSE MODELS OF UNSTABLE DNA

In order to provide a system in which the factors effecting repeat stability may be more precisely defined, a number of mouse models containing expanded CTG•CAG repeat arrays have been generated. These include models for a number of loci and with varying amounts of flanking DNA in simple transgenics,[11-19] and more recently recombination into the murine homologue.[20-22] To date, two models using transgenes derived from the human *DM1* locus have been reported[14,16] (Fig. 1A). The *Dmt*162 transgene contains 162 CTG•CAG repeats, but only incorporates ~700bp of the *DM1* flanking region and none of the coding DNA.[14] Five transgenic mouse lines with random genomic insertion sites were generated with this construct, all of which were unstable in the germline. Although the rates of germline mutation were high, up to 70%, the length change events observed were relatively small, usually less than +/- 10 repeats. Sex-specific differences were observed with expansions predominating in transmissions from males and contractions predominating in female transmissions. In addition, there were dramatic position effects with mutation rates varying from 10% to 70% dependent on the integration site. In order to determine if sequences flanking the human DM1 repeat might be necessary to replicate the human dynamics of the repeat, mice have also been generated using much larger cosmid constructs.[16] The cosmid construct used spanned 45 kb of the DM1 region and includes the upstream gene *DMWD* in addition to *DMPK* and *SIX5*. Multiple lines have been generated with 20 (DM20), 55 (DM55) and 300 (DM300) CTG•CAG repeats.[16,23] Not surprisingly, the repeat in the DM20 lines was very stable with no mutations observed during germline transmission. The repeat was moderately unstable (germline mutation rates 0 to 3%) in the DM55 lines and biased toward small expansions (mostly +1 repeats). The repeat was dramatically more unstable in the germline of the DM300 mice with mutation rates approaching 100%. These were again mostly biased toward expansions (~90%). The length changes observed were much larger than have been observed in other murine models; mean length change +9 repeats in males and +20 repeats from females. However, these length changes are still an order of magnitude smaller than would be expected for similar sized alleles at the human *DM1* locus.[3,5] Nonetheless, the length changes observed in both the *Dmt*162 and DM55 and DM300 lines are comparable to the dynamics observed at many of the more stable human loci such as the spinocerebellar ataxia type 3, spinal and bulbar muscular atrophy and dentatorubral pallidoluysian atrophy loci.[24] Thus, it currently remains unclear whether the large germline expansions observed at the human *DM1* locus are reproducible in mice. Failure so far might be due to the omission of critical *cis*-acting sequences in transgene constructs, the inability of critical *cis*-acting human sequences to mouse genome. Alternatively, the effect may simply reflect the short life cycle of the mouse. Transmitted expansion sizes have been shown to be age dependent in a number of the mouse models,[15,18,23,25] and if the length changes observed after one to two years in mice were extrapolated to the 20 to 40 year reproductive age of humans then they would indeed be comparable to even the most unstable human loci such as *DM1*.

These lines have also been used to determine if somatic instability can be replicated in the mouse. In contrast to the failure to fully replicate germline instability, somatic instability appeared to be highly reproducible in the mouse. Somatic mosaicism in *Dmt*162, DM55 and DM300 mice was expansion-biased, tissue-specific and age-dependent.[23,26,27] Moreover, the length changes observed were large with some cells in *Dmt*162 and DM300 mice containing additional expansion of more than 200 repeats. Most interestingly, the degree and precise tissue specificity of repeat instability was highly line-specific. Only one of the five *Dmt*162 lines displayed significant levels of somatic instability, the remainder remaining very stable throughout life. All of the DM300 lines showed somatic instability, but the absolute tissue specificity differed between them despite the large amount of human genomic DNA that is incorporated. Overall these data suggest that local sequence context may influence the general degree of instability, but that larger scale effects may moderate the tissue specificity. Consistent with these predominantly position-mediated effects, no association with tissue-specificity and cell turnover have been observed casting doubt on the predominant replication slippage based mechanism of DNA instability. Moreover, no association with transcriptional levels and the tissue-specificity have been observed either.[26] However, lines in which the transgenes are not expressed at all appear to the most stable.[15] These data suggest that the genomic environment consistent with gene expression is necessary, but not sufficient, to facilitate somatic instability.

Mouse models generated to understand some of the other human expansion disease loci[15,17,18,20-22] similarly recreate the somatic mosaicism observed in the Dmt162, DM55 and DM300 mice. However, the absolute levels of variation reported are not generally as high. This may reflect transgene content, integration site or allele length effects. More likely however, is a failure to use the sensitive single molecule PCR approaches[5] to detect variation that have been used so successfully in the Dmt162 and DM55 mice.[14,26] Very recently, these methods have been used to reveal gross expansions in the striatum of Huntington disease (HD) knock-in mice.[28] These data suggest that somatic mosaicism may, as with DM1, contribute to the tissue specificity and progressive nature of some the other repeat expansion disorders.

Mice models such as these should prove excellent for further defining the critical factors involved in regulating repeat dynamics. Indeed, transgenics incorporating exon 1 of the human HD gene with an expanded CAG•CTG repeat have been used to reveal that the mismatch repair gene MSH2 is actually required for the development of high levels of somatic mosaicism.[29] These data shed further doubt on the predominant replication slippage model which would predict that loss of mismatch repair activity would actually increase instability. No doubt further studies such as these using the array of mouse DNA repair variants that are now becoming available will shed further light on the role of these genes in the expansion process.

MOLECULAR PATHOGENESIS OF DM1

The DMPK Protein is a Novel Kinase

The CTG expansion in DM1 patients is located within the 3' UTR of a novel protein kinase, named DMPK. After the DM1 gene was cloned, major efforts were focused on the expression analysis of DMPK protein in normal tissues and in DM1 patients and on the identification of the biological function of DMPK.

DMPK protein is expressed in many tissues with highest expression in skeletal muscle and heart and localized to the neuromuscular junction.[30,31] The protein consists of several domains, including an N-terminal leucine-rich region, a catalytic kinase

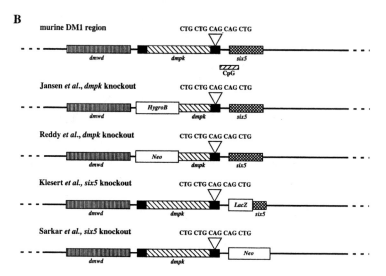

Figure 1B. The mouse myotonic dystrophy type 1 genomic region and replacement alleles. Shown are the mouse *DM1* region including the upstream *DMWD* gene, *Dmpk* and the downstream *Six5* gene, the position of the cryptic CTG repeat and the nearby CpG island. Also shown are the two *Dmpk* and *Six5* replacement alleles: Jansen et al, replaced exons 1-7 of *Dmpk* with a hygromycin B cassette (HygroB) (41); Reddy et al, replaced exons 1-7 of *Dmpk* with a neomycin cassette (Neo) (42); Klesert et al, replaced exons 1 and 2 of *Six5* with a β-galactosidase cassette (LacZ) (54); and Sarkar *et al.*, replaced the whole of *Six5* with a neomycin cassette (53).

domain, a C-terminal coiled-coil domain and a membrane association domain. Analyses of biological function of DMPK domains showed that the kinase domain is required for phosphorylation of serines and threonines in substrate molecules,[32,33] the coiled-coil domain is necessary for DMPK oligomerization[34] and the membrane association domain is involved in peripheral membrane association of the kinase.[35]

Little is known about the role of DMPK in signal transduction pathways. Several molecules have been proposed as candidates for physiological regulatory factors of DMPK. Since DMPK-related proteins are regulated by Rho family GTPases, small G proteins are considered potential DMPK activators.[34] It has been shown that DMPK interacts with Rac-1,[36] a protein that belongs to the Rho family. Because members of the Rho family are associated with the actin cytoskeleton and regulate its dynamic interaction with the plasma membrane, DMPK might participate in the regulation of adhesion-dependent pathways.[36]

One of the most important investigations of DMPK function is identification of its biological substrates. Since myotonia is associated with defects in ion channels, it was suggested that DMPK might be involved in the phosphorylation of ion channels that would affect their function. It was shown that DMPK phosphorylates the β-subunit of voltage-dependent Ca^{2+}-release channel in vitro.[33] In agreement with these data, Ca^{2+} homeostasis was found to be affected in mutant mice deficient for DMPK.[37] Analysis of these mice also showed alterations of activity for Na^+ channels.[38] Recently, it was shown that DMPK phosphorylates phospholemman, a membrane protein that induces Cl^- currents.[39]

Initial hypotheses suggested that DMPK expression might be affected by the CTG expansion in the 3' UTR of the DMPK gene. Immunoanalysis of DMPK protein showed that in a majority of patients, DMPK levels were reduced. However, there are several cases where DMPK protein levels are unchanged or even elevated.[40] In order to understand whether alterations of DMPK expression are crucial for the disease phenotype, mouse models where the *DMPK* gene has been deleted or overexpressed were generated.

Mouse Models of DMPK Function

DMPK is expressed in many tissues in both man and mouse, but is particularly highly expressed in skeletal muscle and heart.[30,31] This observation in addition to the location of the CTG•CAG repeat expansion within the transcriptional unit of the gene, made DMPK a prime candidate for mediating the pathogenicity of the DM1 expansion. A mouse model over-expressing human *DMPK* (Fig. 1B) does show a mild hypertrophic cardiomyopathy and an unexplained increase in neonatal mortality, but no obvious correlation with any of the symptoms observed in DM1 patients.[41] However, two independent knockouts of mouse *Dmpk* (Fig. 1B) have not produced a dramatic phenotype either.[41,42] Even mice completely deficient for *Dmpk* are fully viable and appear morphologically normal. Detailed investigations have revealed some subtle effects. Very old mice do develop a mild skeletal muscle myopathy, but the histological changes and myotonia characteristic of DM1 patients were not reproduced. A convincing cardiac conduction defect has been reported which is similar in nature to that observed in DM1 patients.[42-44] Moreover this effect is observed in mice both homo- and heterozygous for the null allele. *Dmpk* knockout mice developed prolonged AV conduction times and moreover, this defect was age-dependent. There were no cardiac AV problems in 2 months old mice, but they were observed in older mice (> 5 months). Homozygous *Dmpk* knock out mice showed second- and third-degree AV blocks that were absent in heterozygous or wild type mice. There were no differences in conduction defects in *Dmpk* homozygous mice in different age groups. Since the same mice have not shown atrophy, fatty replacements and fibrosis, it seems likely that the conduction defects in patients with DM1 might be associated with the lack of Dmpk, but degeneration of the conduction system might be due to other causes. It is also very interesting, that overexpression of DMPK in mice resulted in the development of hypertrophic cardiomyopathy. These data suggest that DMPK directly or indirectly is involved in the development of cardiac defects.

There are several possible hypothetical explanations how the lack or induction of DMPK would affect heart function. One possibility is that DMPK might regulate specific ion channels that might be affected in DM1 hearts due to abnormal levels of

DMPK kinase. In agreement with this suggestion, Ca^{2+} homeostasis has been reported to be defective in *Dmpk -/-* skeletal muscle cells, although the pathophysiological consequences of this effect are unclear.[37] Alterations in skeletal muscle sodium channel function have also been reported in *Dmpk +/-* mice.[38]

Defects in other organ systems commonly affected in DM1 patients such as the eye, smooth muscle and reproductive tract have not been reported in *Dmpk* deficient mice. Thus, although Dmpk appears to be essential for correct functioning of skeletal and cardiac muscle cells, its absence does not appear to contribute significantly to many of the major features associated with DM1 in humans.

DEFICIENCY OF SIX5 IN DM1

There are several genes in the region surrounding the CTG repeat at the DM1 locus[45] (Fig. 1B), which led to the hypothesis that the expanded repeat might alter the expression of genes in addition to *DMPK*. Repetitive elements in other areas of the genome, for example at the heterochromatin of telomeres and centromeres, were known to suppress the expression of adjacent genes. For example, studies in Drosophila and other organisms demonstrated that genes positioned adjacent to regions of heterochromatin had an increased probability of being suppressed, sometimes resulting in a variegated expression pattern, termed position effect variegation (PEV). Therefore, it seemed plausible that the repetitive sequence introduced at the DM1 locus with CTG repeat expansion might alter the expression of adjacent genes. This hypothesis was indirectly supported by in vitro studies that demonstrated a high affinity of nucleosomes for the CTG sequence,[46] and subsequently by the in vivo demonstration that the region surrounding the expanded repeat had a more condensed chromatin structure than the wild-type allele.[47]

The promoter for the *SIX5* gene, formerly termed Dystrophia Myotonia Associated Homeodomain Protein (*DMAHP*), is very close to the repeat and within the region that exhibits expansion induced changes in chromatin structure. The expansion of the CTG repeat does suppress expression of *SIX5*, since studies in cells from individuals with DM1 demonstrated decreased steady-state levels of the *SIX5* transcript.[48,49] *SIX5* belongs to a family of homeobox transcription factors related to the *Drosophila sine oculis* gene.[50,51] In *Drosophila*, *sine oculis* is part of a group of genes critical for eye development. The same network of genes has been conserved in vertebrates, but as multigene families. The family of vertebrate homologues to the *Drosophila sine oculis* are referred to as the *SIX* genes, and *SIX5* is the family member at the DM1 locus. Different *SIX* gene family members are expressed in many different cell types during vertebrate development, including the vertebrate eye and lens, as well as in skeletal muscle.

Although much remains to be learned regarding the role of the *SIX* family of genes in vertebrate development, the little that is known suggests that they might have a role in the pathogenesis of DM1. Studies of gene promoter elements indicated that the *SIX* family members are important transcription factors for a subset of genes expressed in skeletal muscle and for the expression of subunits of the sodium-

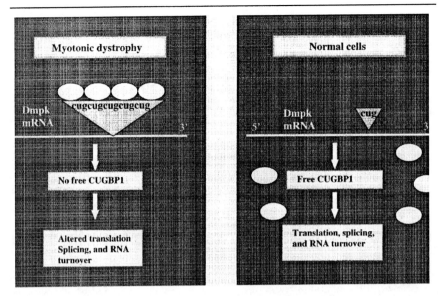

Figure 2. RNA model for DM1 disease. In DM1 patients, CUG repeat is expanded within the DMPK mRNA. CUGBP1 (shown as open oval) is sequestered by expanded CUG repeats. As a result of this sequestration, DM1 cells are lacking of free protein that affects RNA processing. In contrast, in normal cells CUG repeat is not expanded and CUGBP1 protein is free for regulation of RNA processing.

potassium ATPase.[50,52] Abnormalities of sodium homeostasis have been reported in DM1 and could contribute to the myotonia, the cataracts, the cardiac conduction defects, and central nervous system effects. Therefore, the possible roles in skeletal muscle gene expression and in sodium homeostasis make the *SIX5* gene a good candidate regulating some of the critical features of DM1.

As an initial test of the role of the *SIX5* gene in human biology, homologous recombination was used to disrupt the murine *Six5* gene (Fig. 1B). Mice with a deficiency of *Six5* developed cataracts at a young age, strongly suggesting that human *SIX5* deficiency might be the cause of the cataracts associated with DM1.[53,54] It remains possible that other features of DM1 might also be attributed to decreased SIX5 expression, perhaps acting together with a deficiency of DMPK or with a possible gain-of-function role of the CUG repeat in the RNA.

ALTERATIONS OF RNA METABOLISM IN DM1

Given the lack of an overt DM1 phenotype in *Dmpk* knock out mice, several new hypotheses have been suggested. Among those, an RNA based model has been proven by a number of recent publications. It was initially shown that the levels of DMPK mRNA in patients with DM1 were unchanged within total RNA; however, mRNA levels were significantly reduced within poly(A)+mRNA.[55,56] Moreover, it has been shown that the levels of poly(A)-containing DMPK mRNA was reduced

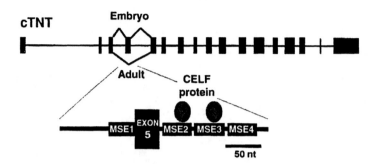

Figure 3. CUGBP1 and ETR-3 like factors (CELF proteins) bind cTnT intronic elements (MSEs) that promote exon inclusion in embryonic muscle.

not only from the mutant allele, but also from a normal allele, suggesting that the CTG expansion has a negative effect on the *DMPK* gene in trans. These data provided a background for a hypothetical RNA model for DM1 disease where trinucleotide repeats might affect genes via association with RNA-binding proteins.[55] This suggestion was further supported by the demonstration that DMPK mutant transcripts formed foci within nuclei of DM1 patients.[57] Similar foci were identified in cultured cells transfected with CUG-expressing constructs.[58] Although the nature of these foci is currently unknown, their ability to hybridize with CAG triplet repeat probes suggests that they are formed by mutant DMPK transcripts.[59]

The RNA-based hypothesis for DM1 pathogenesis has been recently proven by generation of mouse models expressing large expanded CUG repeats. The simple *Te162* transgene that does not express the *DMPK* coding region but expresses a large CUG repeat tract within the *DMPK* 3'UTR has been used to reproduce the testicular atrophy associated with DM1 males[60] (Fig. 1A). More recently, a CUG repeat array has been incorporated into the 3'UTR of a human skeletal muscle α-actin transgene[61] (Fig. 1A). Expression of a 5 CUG repeat allele had no effect in mice, whereas the expression of a large 250 CUG repeat array resulted in muscular atrophy and myotonia: typical characteristics of DM1 patients. These results provide strong evidence for a major role of RNA CUG repeats in the molecular pathogenesis of DM1 and provide an excellent basis for further determining how the effect of RNA CUG repeats is mediated.

The RNA-based model suggests that the expansion of CUG RNA repeats in DM1 alters (sequesters) specific RNA binding proteins that interact with CUG repeats (Fig. 2). Currently, several RNA-binding proteins are considered as candidate factors sequestered by CUG expansion within DMPK mRNA. This group of proteins includes two distinct protein families: CUGBP1-like proteins and EXP (expansion binding) proteins.

Table 1. A family of CUGBP1 and ETR-3 like factors (CELF) bind to bruno element and regulate cTnT splicing

CELF[a]	BRUNOL[b]	% similarity w/ CUGBP1	Binding bruno[b]	cTNT[a]	Splicing activity (cTNT)[a]
CUGBP1	BRUNOL2	----	+	+	+
ETR-3	BRUNOL3	78%	+	+	+
CELF3	BRUNOL1	44%	ND	ND	+
CELF4	BRUNOL4	42%	ND	+	+
CELF5	BRUNOL5	39%	ND	ND	+

a (Ladd et al. 2001); b (Good et al. 2000); ND-not determined

CUGBP1 (CUG RNA-Binding Protein) is Affected in Patients with DM1

An initial search for CUG RNA-binding proteins identified two RNA-binding proteins that specifically interact with CUG_8 repeats.[62] One of these proteins, named ss-CRRP, interacts with single-stranded DNA containing CTG repeats as well as with RNA CUG triplet repeats; while another protein, CUGBP1, interacts only with RNA CUG repeats. Comparison of the CUG-binding activity for CUGBP1 and ss-CRRP in normal individuals and in individuals affected with DM1 showed that binding activity for ss-CRRP is unaltered in DM1; however, the binding activity of CUGBP1 is significantly altered.[63] Because of this finding, CUGBP1 has been further investigated in detail.

Investigations of CUGBP1 in DM1 patients showed that there are significant disease-associated alterations in protein levels, activity and intracellular distribution of CUGBP1. The level of hypophosphorylated CUGBP1 is increased within nuclei of DM1 patients.[64] It has been recently shown that alterations of CUGBP1 expression in DM1 are, at least in part, due to sequestration of CUGBP1 by CUG repeats within the mutant DMPK transcripts[65] (Fig. 2). Sequestration analysis shows that in addition to CUGBP1, an unknown RNA-binding protein of high molecular weight is involved in heavy RNA-protein complexes formed by expanded CUG repeats, suggesting that mutant DMPK mRNA affects more than one RNA-binding protein. Examination of RNA processing in DM1 tissues demonstrated that two levels of RNA processing are affected by alterations in CUGBP1 expression. Analysis of cardiac troponin T (cTnT) alternative splicing in DM1 heart tissue and skeletal muscle cultures demonstrated that alterations in CUGBP1 led to aberrantly high levels of exon inclusion.[66] Splicing of cTnT minigenes in DM1 skeletal muscle cultures indicated the same aberrant pattern as in DM1 patients compared to splicing in skeletal muscle cultures from unaffected controls. Importantly, that aberrant splicing requires a CUGBP1 binding site within the intronic muscle-specific enhancer (MSE), demonstrating that the aberrant splicing is likely to be mediated by CUGBP1 and/or other members of this family (Fig. 3).

Investigations of the effect of RNA CUG repeats on CUGBP1 in cultured cells confirmed that CUG repeats alter CUGBP1 expression and suggested a putative mechanism of this effect. Analysis of RNA-CUGBP1 complexes showed that the majority of CUGBP1 is bound to the endogenous RNA containing CUG repeats in DM1 heart tissue.[65] Similar sequestration of CUGBP1 has been observed in DM1 cell culture models when cells were transfected with plasmid expressing long CUG repeats.[65] Analysis of RNA-CUGBP1 complexes in DM1 cells demonstrated that these complexes contain transcripts with CUG repeats,[65] suggesting that CUGBP1 is sequestered by DMPK mRNA. In vivo data suggest that CUGBP1 binds to long CUG expansions and that this binding leads to the stabilization of CUGBP1.[65] In addition, detailed study of cultured cells expressing RNAs with long expansions (480-1440 CUG repeats) shows that CUGBP1 activity is affected by long CUG repeat sequences, with CUGBP1 activity increasing proportionally to the number of repeats.[66] These data suggest that, similar to tissue culture, CUGBP1 is also sequestered by CUG expansion in DM1 patients. It is interesting to note that electron microscopy studies indicated that, under specific in vitro conditions, CUGBP1 preferentially binds to the single-stranded base of double-stranded hairpin structures that are formed by CUG repeats.[67] This observation offers the possibility that CUGBP1 can be involved in stabilization/destabilization of secondary structures of RNA containing CUG RNA repeats.

CUGBP1 Belongs to a Conserved Family of Elav RNA-Binding Proteins

Comparison of the nucleotide sequence of CUGBP1 with known RNA binding proteins showed a high level of homology to *elav* (embryonic lethal abnormal visual phenotype) family proteins.[68] *Elav* proteins are involved in the regulation of a specific sub-class of mRNAs coding for proteins regulating the cell cycle. For example, the binding of *elav* proteins to c-myc or c-fos mRNAs affects mRNA stability or translation and this leads to alteration of protein levels affecting overall proliferative cellular status. *Elav* proteins in Drosophila are located in nuclei where they regulate splicing. In contrast, in human cells elav proteins are located in both cytoplasm and nuclei and are involved in multiple steps of RNA processing such as stability and translation. Similar to the *elav* proteins, CUGBP1 contains three RNA binding domains (RBDs), the distribution of which within CUGBP1 is similar to that observed in *elav*-like proteins—the first two RBDs are located close to each other, but RBDIII is separated from the first two RBDs by a long linker.[68] It has been suggested that separation of RBD1+2 and RBD3 might be associated with two distinct biological functions of RNA-binding proteins and with different sequence specificity.

CUGBP1 TARGETS

Since CUGBP1 is affected in DM1 patients, identification of its native mRNA targets is important for understanding CUGBP1 downstream pathways. So far, two RNAs regulated by CUGBP1 have been characterized in detail. They include premRNA coding for cardiac Troponin T (cTnT)[67] and the mRNA for a transcription factor CCAAT/Enhancer Binding Protein β, C/EBPβ.[69] CUGBP1 binds to CUG repeats within cTnT premRNA and regulates splicing of a single alternative exon that is included in embryonic striated muscle and skipped in the adult.[67,70] Exon inclusion in embryonic striated muscle requires four intronic muscle-specific enhancers (MSEs) located upstream and downstream of the alternative exon (Fig. 3). These elements are necessary and sufficient to promote exon inclusion of a heterologous exon in embryonic striated muscle. CUGBP1 binds directly to the conserved MSEs and promotes inclusion of an alternative exon. Mutations in the MSEs that prevent CUGBP1 protein binding also prevent activation of exon inclusion by exogenous CUGBP1.[66] Other alternatively spliced premRNAs potentially regulated by CUGBP1 or CUGBP1 homologous proteins include the neuron-specific and developmentally regulated exon 82 of GABA$_A$. A CUGBP1 binding site has been mapped within the intron immediately upstream of the exon.[71] Additionally, coexpression of CUGBP1 with an amyloid precursor protein (APP) minigene increased exon skipping of exon 8,[72] suggesting that APP splicing might be regulated by CUGBP1 or CUGBP1 family members. The contribution of genes with disrupted splicing in DM1 pathology remains to be determined.

Significant amounts of CUGBP1 have been detected in cytoplasm, suggesting that CUGBP1 is involved in processing of RNAs in the cytoplasm as well as in nuclei. Investigations of the binding of CUGBP1 to a number of mRNAs showed that CUGBP1 binds to the 5'region of mRNA coding for the transcription factor C/EBPβ. A single C/EBPβ mRNA produces several protein isoforms (full-length protein and two truncated isoforms—liver inhibitor protein, LIP, and liver activator protein, LAP) via alternative initiation from downstream AUG codons.[73] It has been found that CUGBP1 binds to the 5' region of C/EBPβ mRNA and induces translation of the dominant negative molecule LIP.[69] In agreement with observations obtained in tissue culture systems,[65] an increase of CUGBP1 binding activity in DM1 patients also results in induction of LIP.[65] Since overexpression of the LIP isoform alters cell proliferation,[74] the increase of LIP levels in patients with DM1 suggests that proliferation rate might be also affected in DM1 disease.

OTHER MEMBERS OF CUGBP1 FAMILY

CUGBP1 is a member of a family of proteins called CUGBP1-like proteins or CELF proteins (CUGBP1 and ETR-3 like factors).[70] This family also includes three other proteins called CELF3, CELF4, and CELF5.[70] The CUGBP1 family has also been called BRUNOL because of their homology with the Drosophila bruno protein[75]

(Table 1). CUGBP1-like proteins are expressed in a tissue specific manner. For example, CUGBP1 is widely expressed with high levels in skeletal muscle and heart,[76] and ETR-3 is expressed in heart[77] as well as in brain and striated muscle.[70] This family is likely to function in multiple aspects of RNA processing and translation. CUGBP1 is closely related to the EDEN-binding protein (EDEN-BP) in Xenopus as well as bruno in Drosophila. Both of these proteins regulate translation by the interaction with specific elements within the 3' UTR of target mRNAs.[78,79] ETR-3 protein was originally identified within human heart,[80] and it is identical to a recently identified protein named apoptosis-related protein (APRP). APRP was identified as a differentially expressed gene in human neuroblastoma.[81] Human ETR-3/APRP is abundant in cardiac tissue, suggesting that it might be involved in the regulation of cardiac specific mRNAs. Analysis of RNA-binding activity of ETR-3 showed that, similar to CUGBP1, it binds to CUG repeats.[77] It has been shown that ETR-3 is also capable of regulating the alternative splicing of cTnT[70] and APP.[72] It remains to investigate whether ETR-3 and other CUGBP1 homologous proteins are affected in DM1 patients. Since these proteins are expressed in a tissue specific manner, it is possible that different members of this family function in different tissues, inducing tissue specific symptoms in DM1 disease.

EXP/MNBL Represent a Second Family of CUG Repeat Binding Proteins

Recently, another family of CUG binding proteins has been identified.[82] Sequence analysis of these proteins, which bind to double-stranded RNA CUG repeats, indicated that they are homologous to the Drosophila muscleblind protein.[83] In Drosophila, muscleblind protein is required for myogenic and photoreceptor differentiation suggesting that EXPs in humans might be involved in skeletal muscle and eye development. In vitro analysis of EXP proteins by UV-cross link assay demonstrated that EXPs bind efficiently to expanded CUG repeat sequences.[82] Although it is unknown whether EXP proteins also bind to long CUG expansion in vivo, these proteins could potentially be affected in DM1 by expansion of CUG repeats within the mutant DMPK mRNA. In agreement with this suggestion, immunofluorescence analysis of EXP in DM1 muscle cell line indicated formation of foci within DM1 myoblast nuclei.[82] Further studies are required to investigate the function of EXPs, to find their native targets, and to determine DM1 symptoms that are associated with EXPs.

CONCLUSIONS

DM1 is one of the most complex diseases both at the clinical and molecular levels. Discovery of a CTG/CUG unstable expansion in the 3' UTR of the DM1 gene prompted researchers to investigate the biological effects of untranslated unstable elements on the structure of chromatin, efficiency of gene transcription,

RNA processing, and signal transduction pathways. These studies provided knowledge that some of the main features of DM1 such as myotonia and testicular atrophy are due to expansion of RNA CUG repeats, while cardiac abnormalities and cataracts are associated with *DMPK* and *SIX5* genes respectively. While the details of each mechanism are being investigated, development of therapy is underway. For example, a trans-splicing ribozyme is able to shorten the CUG triplet repeat expansion within mutant RNA and repair it.[84] Therefore, application of ribozymes for DM1 therapy is a perspective strategy to correct the dominant-negative effect of CUG repeats. Additional studies are required to understand the interaction and overlaps between pathological pathways induced by CTG/CUG repeat expansion in patient tissues.

ACKNOWLEDGMENTS

The author's research is supported by grants from National Institutes of Health RO1AR44387 (LTT), RO3AG16392 (LTT), RO1AR45203 (SJT), RO1AR45653 (TAC) and grants from Muscular Dystrophy Association (LTT and TAC). DGM is a Lister Institute Research Fellow.

REFERENCES

1. Harper PS. Myotonic dystrophy and other autosomal muscular dystrophies. In: Scriver CR, Beaudet AL, Sly WS, Valle D, eds. The Metabolic and Molecular Bases of Inherited Diseases. 7th ed. New York: McGraw-Hill, Inc., 1995:4227-4251.
2. Harper PS, Harley HG, Reardon W et al. Anticipation in myotonic dystrophy: New light on an old problem. Am J Hum Genet 1992; 51:10-16.
3. Lavedan C, Hofmann-Radvanyi H, Shelbourne P et al. Myotonic dystrophy: Size- and sex-dependent dynamics of CTG meiotic instability, and somatic mosaicism. Am J Hum Genet 1993; 52:875-883.
4. Tsilfidis C, MacKenzie AE, Mettler G et al. Correlation between CTG trinucleotide repeat length and frequency of severe congenital myotonic dystrophy. Nat Genet 1992; 1:192-195.
5. Monckton DG, Wong L-JC, Ashizawa T et al. Somatic mosaicism, germline expansions, germline reversions and intergenerational reductions in myotonic dystrophy males: Small pool PCR analyses. Hum Mol Genet 1995; 4:1-8.
6. Wong L-JC, Ashizawa T, Monckton DG et al. Somatic heterogeneity of the CTG repeat in myotonic dystrophy is age and size dependent. Am J Hum Genet 1995; 56:114-122.
7. Martorell L, Monckton DG, Gamez J et al. Progression of somatic CTG repeat length heterogeneity in the blood cells of myotonic dystrophy patients. Hum Mol Genet 1998; 7:307-312.
8. Ashizawa T, Dubel JR, Harati Y. Somatic instability of CTG repeat in myotonic dystrophy. Neurology 1993; 43:2674-2678.
9. Anvret M, Ahlberg G, Grandell U et al. Larger expansions of the CTG repeat in muscle compared to lymphocytes from patients with myotonic dystrophy. Hum Mol Genet 1993; 2:1397-1400.
10. Thornton CA, Johnson KJ, Moxley RT. Myotonic dystrophy patients have larger CTG expansions in skeletal muscle than in leukocytes. Ann Neurol 1994; 35:104-107.
11. Bingham PM, Scott MO, Wang S et al. Stability of an expanded trinucleotide repeat in the androgen receptor gene in transgenic mice. Nat Genet 1995; 9:191-196.

12. Burright EN, Clark HB, Servadio A et al. SCA1 transgenic mice: A model for meurodegeneration caused by an expanded CAG trinucleotide repeat. Cell 1995; 82:937-948.
13. Goldberg YP, Kalchman MA, Metzler M et al. Absence of disease phenotype and intergenerational stability of the CAG repeat in transgenic mice expressing the human Huntington disease transcript. Hum Mol Genet 1996; 5:177-185.
14. Monckton DG, Coolbaugh MI, Ashizawa K et al. Hypermutable myotonic dystrophy CTG repeats in transgenic mice. Nat Genet 1997; 15:193-196.
15. Mangiarinin L, Sathasivam K, Mahal A et al. Instability of highly expanded CAG repeats in mice transgenic for the Huntington's disease mutation. Nat Genet 1997; 15:197-200.
16. Gourdon G, Radvanyi F, Lia AS et al. Moderate intergenerational and somatic stability of a 55 CTG repeat in transgenic mice. Nat Genet 1997; 15:190-192.
17. La Spada AR, Peterson KR, Meadows SA et al. Androgen receptor YAC transgenic mice carrying CAG 45 alleles show trinucleotide repeat instability. Hum Mol Genet 1998; 7:959-967.
18. Sato T, Oyake M, Nakamura K et al. Transgenic mice harboring a full-length human mutant DRPLA gene exhibit age-dependent intergenerational and somatic instabilities of CAG repeats comparable with those in DRPLA patients. Hum Mol Genet 1999; 8:99-106.
19. Ikeda H, Yamaguchi M, Sugai S et al. Expanded polyglutamine in the Machado-Joseph disease protein induces cell-death in vitro and in vivo. Nat Genet 1996; 13:196-202.
20. Wheeler VC, Auerbach W, White JK et al. Length-dependent gametic CAG repeat instability in the Huntington's disease knock-in mouse. Hum Mol Genet 1999; 8:115-122.
21. Shelbourne PF, Killeen N, Hevner RF et al. A Huntington's disease CAG expansion at the murine Hdh locus is unstable and associated with behavioral abnormalities in mice. Hum Mol Genet 1999; 8:763-774.
22. Lorenzetti D, Watase K, Xu B et al. Repeat instability and motor incoordination in mice with a targeted expanded CAG repeat in the Sca1 locus. Hum Mol Genet 2000; 9:779-785.
23. Seznec H, Lia-Baldini AS, Duros C et al. Transgenic mice carrying large human genomic sequences with expanded CTG repeat mimic closely DM CTG repeat intergenerational and somatic instability. Hum Mol Genet 2000; 9:1185-1194.
24. Brock GJR, Anderson NH, Monckton DG. Cis-acting modifiers of expanded CAG/CTG triplet repeat expandability: Associations with flanking GC content and proximity to CpG islands. Hum Mol Genet 1999; 8:1061-1067.
25. Kaytor MD, Burright EN, Duvick LA et al. Increased trinucleotide instability with advanced maternal age. Hum Mol Genet 1997; 6:2135-2139.
26. Lia AS, Seznec H, Hoffman-Radvanyi H et al. Somatic instability of the CTG repeat in mice transgenic for the myotonic dystrophy region is age dependent but not correlated to the relative intertissue transcription levels and proliferative capacities. Hum Mol Genet 1998; 7:1285-1291.
27. Fortune MT, Vassilopoulos C, Coolbaught MI et al. Dramatic, expansion-biased, age-dependent, tissue-specific somatic mosaicism in a transgenic mouse model of triplet repeat instability. Hum Mol Genet 2000; 9:439-445.
28. Kennedy L, Shelbourne PF. Dramatic mutation instability in HD mouse striatum: Does polyglutamine load contribute to cell-specific vulnerability in Huntington's disease? Hum Mol Genet 2000; 9:2539-2544.
29. Manley K, Shirley TL, Flaherty L et al. MSH2 deficiency prevents in vivo somatic instability of the CAG repeat in Huntington disease transgenic mice. Nat Genet 1999; 23:471-473.
30. Whiting EJ, Waring JD, Tamai K et al. Characterization of myotonic dystrophy kinase (DMK) protein in human and rodent muscle and central nervous tissue. Hum Mol Genet 1995; 4:1063-1072.
31. Lam LT, Pham YC, Man N et al. Characterization of a monoclonal antibody panel shows that the myotonic protein kinase, DMPK, is expressed almost exclusively in muscle and heart. Hum Mol Genet 2000; 9:2167-2173.

32. Dunne PW, Walch ET, Epstein HF. Phosphorylation reactions of recombinant human myotonic dystrophy protein kinase and their inhibition. Biochem 1994; 33:10809-10814.

33. Timchenko L, Nastainczyk W, Schneider T et al. Full-length Myotonin protein kinase (72 kDa) displays serine kinase activity. Proc Natl Acad Sci (USA) 1995; 92:5366-5370.

34. Bush EW, Helmke SM, Birnbaum A et al. Myotonic dystrophy protein kinase domains mediate localization, oligomerization, novel catalytic activity, and autoinhibition. Biochem 2000; 39:8480-8490.

35. Waring JD, Haq R, Tamai K et al. Investigation of myotonic dystrophy kinase isoform translocation and membrane association. J Biol Chem 1996; 271:15187-15193.

36. Shimizu M, Wang W, Walch ET et al. Rac-1 and Raf-1 kinases, components of distinct signalling pathways, activate myotonic dystrophy protein kinase. FEBS Letters 2000; 475:273-277.

37. Benders AA, Groenen PJ, Oerlemans FT et al. Myotonic dystrophy protein kinase is involved in the modulation of the Ca2+ homeostasis in skeletal muscle cells. J Clin Invest 1997; 100:1440-1447.

38. Mounsey JP, Mistry DJ, Ai CW et al. Skeletal muscle sodium channel gating in mice deficient in myotonic dystrophy protein kinase. Hum Mol Genet 2000; 9:2313-2320.

39. Mounsey JP, John JE III, Helmke SM et al. Phospholemman is a substrate for myotonic protein kinase. J Biol Chem 2000; 275:23362-23367.

40. Hofmann-Radvanyi H, Junien C. Myotonic dystrophy: Over-expression or/and under-expression? A critical review on a controversial point. Neuromuscul Disorders 1993; 3:491-501.

41. Jansen G, Groenen PJTA, Bachner D et al. Abnormal myotonic dystrophy protein kinase levels produce only mild myopathy in mice. Nat Genet 1996; 13:316-324.

42. Reddy S, Smith DBJ, Rich MM et al. Mice lacking the myotonic dystrophy protein kinase develop a late onset progressive myopathy. Nat genet 1996; 13:325-335.

43. Saba S, Vanderbrink BA, Luciano B et al. Localization of the sites of conduction abnormalities in a mouse model of myotonic dystrophy. J Cardiovasc Electrophysiol 1999; 10:1214-1220.

44. Berul CI, Maguire CT, Aronovitz MJ et al. DMPK dosage alterations result in atrioventricular conduction abnormalities in a mouse myotonic dystrophy model. J Clin Invest 1999; 103:1-7.

45. Alwazzan M, Hamshere MG, Lennon GG et al. Six transcripts map within 20 kilobases of the myotonic dystrophy expanded repeat. Mammal Genome 1998:485-487.

46. Wang YH, Amirhaeri S, Kang S et al. Preferential nucleosome assembly at DNA triplet repeats from the myotonic dystrophy gene. Science 1994:669-671.

47. Otten AD, Tapscott SJ. Triplet repeat expansion in myotonic dystrophy alters adjacent chromatin structure. Proc Natl Acad Sci USA 1995; 92:5465-5469.

48. Thornton CA, Wymer JP, Simmons Z et al. Expansion of the myotonic dystrophy CTG repeat reduces expression of the flanking DMAHP gene. Nat Genet 1997; 16:407-409.

49. Klesert TR, Otten AD, Bird TD et al. Trinucleotide repeat expansion at the myotonic dystrophy locus reduces expression of DMAHP. Nat Genet 1997; 16:402-406.

50. Kawakami K, Sato S, Ozaki H et al. Six family genes—Structure and function as transcription factors and their role in development. BioEssays 2000; 22:616-626.

51. Relaix F, Buckingham M. From insect eye to vertebrate muscle: redeployment of a regulatory network. Genes Dev 1999; 13:3171-3178.

52. Heanue TA, Reshef R, Davis RJ et al. Synergistic regulation of vertebrate muscle development by Dach2, Eya2, and Six1, homologs of genes required for Drosophila eye formation. Genes Dev 1999; 13:3231-3243.

53. Sarkar PS, Appukuttam B, Han J et al. Heterozygous loss of Six5 in mice is sufficient to cause ocular cataracts. Nat Genet 2000; 25:110-114.

54. Klesert TR, Cho DH, Clark JI et al. Mice deficient in Six5 develop cataracts: Implications for myotonic dystrophy. Nat Genet 2000:105-109.

55. Wang J, Pegoraro E, Menegazzo E et al. Myotonic dystrophy: Evidence for a possible dominant-negative RNA mutation. Hum Mol Genet 1995; 4:599-606.

56. Krahe R, Ashizawa T, Abbruzzese C et al. Effect of myotonic dystrophy trinucleotide repeat expansion on DMPK transcription and processing. Genomics 1995; 28:1-14.

57. Taneja KL, McCurrach M, Schalling M et al. Foci of trinucleotide repeat transcripts in nuclei of myotonic dystrophy cells and tissues. J Cell Biol 1995; 128:995-1002.

58. Amack JD, Paguio AP, Mahadevan MS. Cis and trans effects of the myotonic dystrophy (DM) mutation in a cell culture model. Hum Mol Genet 1999; 8:1975-1984.

59. Davis BM, McCurrach ME, Taneja KL et al. Expansion of a CUG trinucleotide repeat in the 3' untranslated region of myotonic dystrophy protein kinase transcripts results in nuclear retention of transcripts. Proc Natl Acad Sci USA 1997; 94:7388-7393.

60. Monckton DG, Ashizawa T, Siciliano MJ. Murine models for myotonic dystrophy. In: Wells RD and Warren ST, eds. Genetics Instabilities and Hereditary Neurological Diseases. San Diego: Academic Press, 1998:181-193.

61. Mankodi A, Logigian E, Callahan L et al. Myotonic dystrophy in transgenic mice expressing an expanded CUG repeat. Science 2000; 289:1769-1773.

62. Timchenko LT, Timchenko NA, Caskey CT et al. Novel proteins with binding specificity for DNA repeats and RNA CUG repeats: implications for myotonic dystrophy. Hum Mol Genet 1996; 5:115-121.

63. Timchenko LT, Miller JW, Timchenko NA et al. Identification of a (CUG)n triplet repeat RNA-binding protein and its expression in myotonic dystrophy. Nucl Acids Res 1996; 24:4407-4414.

64. Roberts R, Timchenko NA, Miller JW et al. Altered phosphorylation and intracellular distribution of a (CUG)n triplet repeat RNA-binding protein in patients with myotonic dystrophy and in myotonin protein kinase knockout mice. Proc Natl Acad Sci USA 1997; 94:13221-13226.

65. Timchenko NA, Cai Z-J, Welm AL et al. RNA CUG repeats sequester CUGBP1 and alter protein levels and stability of CUGBP1. J Biol Chem 2001; 276:7820-7826.

66. Phillips AV, Timchenko LT, Cooper TA. Disruption of splicing regulated by a CUG-binding protein in myotonic dystrophy. Science 1998; 280:737-741.

67. Michalowski S, Miller JW, Urbinati CR et al. Visualization of double-stranded RNAs from the myotonic dystrophy protein kinase gene and interactions with CUG-binding protein. Nucl Acids Res 1999; 27:3534-3542.

68. Antic D, Keene JD. Embryonic lethal abnormal visual RNA-binding proteins involved in growth, differentiation, and posttranscriptional gene expression. Am J Hum Genet 1997; 61:273-278.

69. Timchenko NA, Welm AL, Lu X et al. CUG repeat binding protein (CUGBP1) interacts with the 5' region of C/EBPβ mRNA and regulates translation of C/EBP_ isoforms. Nucl Acids Res 1999; 27:4517-4525.

70. Ladd AN, Charlet-BN, Cooper TA. The CELF family of RNA binding proteins is implicated in cell-specific and developmentally regulated alternative splicing. Mol Cell Biol 2001; 21:1285-1296.

71. Zhang L, Liu W, Grabowski PJ. Coordinate repression of a trio of neuron-specific splicing events by the splicing regulator PTB. RNA 1999; 5:117-130.

72. Poleev A, Hartman A, Stamm S. A trans-acting factor, isolated by three hybrid system that influences alternative splicing of the amyloid precursor protein minigene. Eur J Biochem 2000; 267:4002-4010.

73. Descombes P., Schibler U. A liver-enriched transcriptional activator protein, LAP, and a transcriptional inhibitory protein, LIP, are translated from the same mRNA. Cell 1991; 67:569-579.

74. Calhoven CF, Muller C, Leutz A. Translational control of C/EBPalpha and C/EBPbeta isoform expression. Genes Dev 2000; 14:1920-1932.

75. Good PJ, Chen Q, Warner SJ et al. A family of human RNA-binding proteins related to the Drosophila Bruno translational regulator. J Biol Chem 2000; 275:28583-28592.

76. Caskey CT, Swanson MS, Timchenko LT. Myotonic dystrophy: Discussion of molecular mechanism. Cold Spring Harbor Symp Quant Biol 1996; 61:607-614.

77. Lu X, Timchenko NA, Timchenko LT. Cardiac elav-type RNA-binding protein (ETR-3) binds to RNA CUG repeats expanded in myotonic dystrophy. Hum Mol Genet 1999; 8:53-60.

78. Kim-Ha J, Kerr K, Macdonald PM. Translational regulation of oskar mRNA by bruno, an ovarian RNA-binding protein, is essential. Cell 1995; 81:403-412.

79. Paillard L, Omilli F, Legagneux V et al. EDEN and EDEN-BP, a cis element and an associated factor that mediates sequence-specific mRNA deadenylation in Xenopus embryos. EMBO J 1998; 17:278-287.

80. Hwang DM, Hwang WS, Liew CC. Single pass sequencing of a unidirectional human fetal heart cDNA library to discover novel genes of the cardiovascular system. J Mol Card 1994; 26:1329-1333.

81. Choi DK, Ito T, Tsukahara F et al. Developmentally regulated expression of mNapor encoding an apoptosis-induced ELAV-type RNA binding protein. Gene 1999; 237:135-142.

82. Miller JW, Urbinati CR, Teng-Unuay P et al. Recruitment of human muscleblind proteins to (CUG)n expansions associated with myotonic dystrophy. EMBO J 2000; 19:4439-4448.

83. Begemann G, Paricio N, Artero R et al. *muscleblind*, a gene required for photoreceptor differentiation in *Drosophila*, encodes novel nuclear Cys3His-type zinc-finger-containing proteins. Development 1997; 124:4321-4331.

84. Phylactou LA, Darrah C, Wood MJA. Ribozyme-mediated trans-splicing of a trinucleotide repeat. Nat Genet 1998; 18:378-381.

SPINOCEREBELLAR ATAXIAS CAUSED BY POLYGLUTAMINE EXPANSIONS

Giovanni Stevanin,[1,2] Alexandra Dürr[1,2,3] and Alexis Brice[1,2,3]

Autosomal dominant cerebellar ataxias (ADCA) constitute a group of disorders, clinically and molecularly heterogeneous. They are characterized by variable degrees of cerebellar and brainstem degeneration or dysfunction. Neuronal loss variably affects the pons, the inferior olive, the basal ganglia, the cerebellum and its afferent and efferent fibers. Onset is generally during the third or fourth decade but can also occur in childhood or in the old age. Patients usually present with progressive cerebellar ataxia and associated neurological signs, such as ophthalmoplegia, pyramidal or extrapyramidal signs, deep sensory loss and dementia. Attempts to classify subtypes of ADCA were largely unsatisfactory until AE Harding distinguished three phenotypes based on clinical associated signs.[1] ADCA type I is the most common subtype and variably combines cerebellar ataxia, dysarthria, ophthalmoplegia, pyramidal and extrapyramidal signs, deep sensory loss, amyotrophy and dementia. However, several other signs and symptoms may also be associated, i.e., slow eye movements, sphincter disturbances, axonal neuropathy, fasciculations and/or swallowing difficulties. ADCA type II was first described by Froment et al[2] and is characterized by the association of progressive macular degeneration with cerebellar ataxia. Finally, ADCA type III denotes a "pure", generally late onset, cerebellar syndrome.

Molecular genetic studies have revealed that ADCA are also genetically heterogeneous and have led to the mapping of 14 different loci accounting for the disease (for a review, see Refs. 3 and 4). The genes and the responsible mutations have been characterized for most of these loci (Table 1). Polyglutamine-coding $(CAG)_n$ repeat expansions are responsible for the disease in six of these genes denoted spinocerebellar ataxia (SCA) 1-3,[5-9] 6,[10] 7,[11] and 15[12] and account for approximately 40-90% of all ADCA depending on geographical origin. Noncoding repeat expansion

[1]INSERM U289, [2]Institut Fédératif di Recherche des Neurosciences, [3]Département de Génétique, Cytogénétique et Embryologie, Groupe Hospitalier Pitié-Salpêtriére, Paris, France.

Table 1. Clinical[1] and molecular classification of ADCA

Type	Signs associated with cerebellar ataxia	Gene	Locus	Frequency	Mutation	Repeat number range	
						Normal	Pathological
I	Variable	SCA1	6p	0-72%	CAG (coding)	6-44	39-83
	(±ophthalmoplegia	SCA2	12q	4-63%	CAG (coding)	13-33	32-77
	±optic atrophy	SCA3 (MJD1)	14q	0-74%	CAG (coding)	12-47	54-89
	±dementia	SCA4	16q	few	?		
	±extrapyramidal	SCA12 (PP2A)	5	2 families	CAG (non coding)	<29	>65
	signs	SCA14	19q	1 family	?		
	±amyotrophy)	SCA15 (TBP)	6q	9 families	CAG (coding)	27-42	44-63
II	As in type I + progressive macular dystrophy	SCA7	3p	0-16%	CAG (coding)	4-35	36-306
III	Pure cerebellar	SCA5	11cen	2 families	?		
	syndrome	SCA16(CACNA1A)	19p	1-31%	CAG (coding)	4-18	20/21-33
		SCA8	13q	~5%	CTG (non coding)	16-92	107-250
		SCA11	15q	few	?		
Others	Ataxia+epilepsy	SCA10	22q	5 families	ATTCT (non coding)	10-22	>800
	Ataxia+mental retardation	SCA13	19q	1 family	?		

PP2A-PR55β protein phosphatase 2A regulatory sub-unit; CACNA1A voltage-dependent calcium channel sub-unit α1A; MJD1 Machado-Joseph disease 1; TBP tata binding protein

Table 2. Comparison of CAG repeat instability during transmission to progeny.

	Gender of the transmitting parent	
	Male	Female
SCA1	+ 2,0 (-2 to +8, n=16)	+ 0,2 (-1 to +1, n=5)
SCA2	+3,5 (-8 to +17, n=33)	+1,7 (-4 to +8, n=23)
SCA3/MJD	+0,9 (-3 to +5, n=26)	+0,6 (-8 to +3, n=34)
SCA7	+12,1 (0 to +85, n=34)	+4,8 (-6 to +18, n=34)
DRPLA	+7,0 (0 to +28, n=33)	+0,3 (-4 to +4, n=9)
SBMA	+1,8 (-2 to +5, n=11)	+0,2 (-4 to +2, n=20)
HD	+6,1 (-4 to +74, n=156)	+0,6 (-4 to +16, n=160)

From reference 3 by courtesy of the Nature Publishing Group.

has recently been described at the *SCA8*,[13] *SCA10*[14] and *SCA12*[15] loci, but its relevance to the disease remains to be proved. At the *SCA8* locus, the high frequency of large alleles in controls[16-18] makes unlikely that expansion at this locus can cause ataxia alone. Lastly, analysis of large kindreds in which these mutations and loci could be excluded has revealed that other loci must be implicated.[19,20]

POLYGLUTAMINE EXPANSIONS AS MAJOR MUTATIONS IN ADCA

General Characteristics

Large series of patients with various geographical origins have now been reported. Expansions of polyglutamine (polyQ) CAG coding repeats have common properties with three other neurodegenerative diseases carrying the same kind of causative mutation : Huntington's disease (HD), dentatorubro-pallidoluysian atrophy (DRPLA) and spinobulbar muscular atrophy (SBMA):[21]
- onset is mostly in adulthood, but some juvenile cases are observed, especially when transmitted by affected fathers;
- the disease course is progressive, unremitting and usually fatal 10-30 years after onset;
- normal and pathological alleles carry a variable number of CAG repeats, but the clinical symptoms appear above a threshold number of CAG repeats ranging from 20/21 in *SCA6*[22] to 54/55 in *SCA3*/MJD;[23]
- there is a strong negative correlation between the number of CAG repeats and the age at onset;
- the repeat sequence of the pathological alleles is unstable, except for *SCA6*, and its increase in size during transmission results in genetic anticipation;

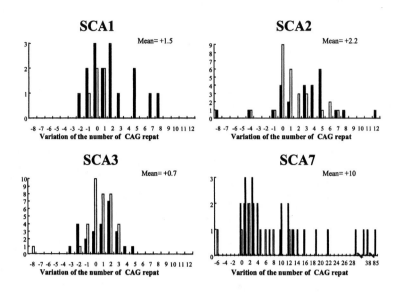

Figure 1. Variation of the number of CAG repeats in spinocerebellar ataxias.[3]

- the genes are expressed ubiquitously; and
- the pathological protein accumulates in neuronal inclusions in several affected but also in nonaffected tissues.

Epidemiology: Prevalence of ADCA and Frequencies of SCAs

ADCA are rare conditions with a prevalence of 1/100,000 in most countries,[24-27] except in populations with probable founder effects, i.e., *SCA2* in the province of Holguin in Cuba (4/10,000)[28] or *SCA3*/MJD in the Azores (1/4000).[29] Founder effects have also been found or suspected at the *SCA1* locus in Japan,[30] *SCA2* in Northern Europe[31] and India,[32] *SCA3*/MJD in France, Portugal and Japan,[33-37] *SCA6* in Germany[38] and *SCA7* in Scandinavia, Korea, North-Africa, Continental Europe and Anglo-Saxon countries.[39,40] This accounts for the variation in the frequency of each *SCA* according to the geographical origin.[41-47] In most countries, however, *SCA3*/MJD is the major locus, with the highest frequency of about 80% in Portugal and the Azores.[43] Its high frequency in other countries probably results from the dissemination of the "Azorean mutations" by Portuguese sailors[29,35] and by the occurrence of independent mutations in distinct populations such as in black Africans and Jewish Yemenites.[37,48] Surprisingly, the *SCA3* mutation was not found in Italy.[42]

The frequency of mutations can also vary greatly in the same country. This is well illustrated in Italy where the *SCA2* mutation accounts for two-thirds of ADCA

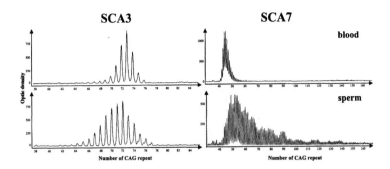

Figure 2. Comparison of CAG repeat expansions in blood (top) and sperm (bottom) DNA of *SCA3* and *SCA7* patients. Electrophoretic profiles of PCR amplified CAG repeats were obtained with Genscan software (Perkin-Elmer). Horizontal axis: number of CAG repeats, vertical axis: fluorescence intensity.

cases in the South and the *SCA1* mutation accounts for three-quarters of the patients in the North.[42]

Genetic Characteristics

Repeat Instability

Normal alleles are usually transmitted to progeny without modification and are similar in size in all tissues. There is, however, a mosaicism of the size of most expansions (except for *SCA6*) that can be visualized in somatic, including the central nervous system, and gonadic tissues.

When visualized in leukocytes, expansions have a tendency to increase in successive generations, the mean ranging from approximately +1 for *SCA3*/MJD to +12 repeats for *SCA7* (Fig. 1 and Table 2). At the *SCA1*, *SCA2* and *SCA7* loci, there is a tendency for greater instability during paternal than during maternal transmissions (Table 2), particularly for the largest expansions (> 20 CAG units).

CAG repeat instability is thought to result from slippage during DNA replication or from the formation of stable hairpin structures.[49,50] However, a recent study in an animal model of HD revealed that the mosaicism in post-mitotic neurons increases with age, suggesting that instability does not solely occur during replication.[51] The differences in instability among polyQ diseases and the increased instability in paternal transmissions indicates that other factors play a role. Instability is influenced by the size of the repeat required to form stable structures, as demonstrated in HD and *SCA7*.[52,53] In *SCA3*/MJD, the analysis of polymorphisms located close to the CAG repeat showed that they act both in trans and in cis.[54,55]

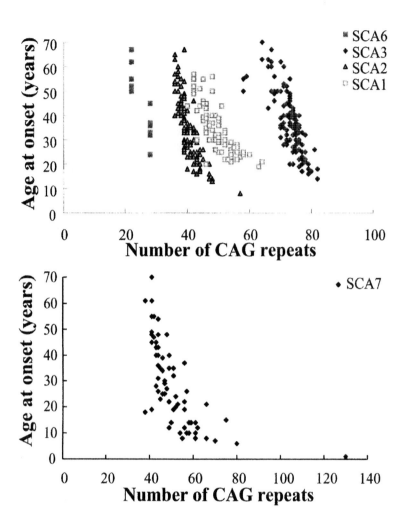

Figure 3. Age at onset/CAG repeat number correlation curves in spinocerebellar ataxias. From reference 3 by courtesy of the Nature Publishing Group.

The greater number of cell divisions in spermatogenesis than in oogenesis is probably another factor. Indeed, mosaicism in gonads[56] is much more pronounced than that observed following an analysis of parent-child transmissions. Whole or single sperm studies revealed much greater mosaicism of the expansion at the *SCA7* locus than at the *SCA3*/MJD locus (Fig. 2), which is in accordance with the differences observed in leukocytes during transmissions.[53,55,57] The massive CAG expansions in *SCA7* may lead to embryonic lethality or dysfunctional sperm,[58] as suggested

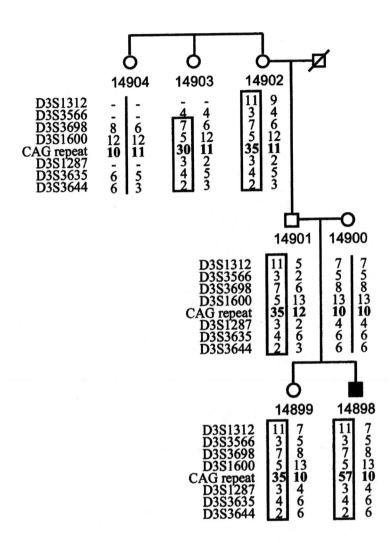

Figure 4. Partial pedigree of a *SCA7* family showing a de novo expansion of the CAG repeat. Haplotypes for several flanking microsatellite markers are shown. Reprinted from reference 203 by courtesy of MARCEL DEKKER, Inc.

initially by the excess of maternal transmissions observed in this disease in *SCA7* kindreds.[59] Meiotic distortion, however, was not observed at the *SCA3* locus.[60]

Table 3. Frequency of neurological signs associated with SCA mutations.[3]
Frequency: 0 = absent; ± = rare; + = 5-24 %; ++ = 25-74 %; +++ = 75-100 %.
From reference 3 by courtesy of the Nature Publishing Group.

	SCA1	SCA2	SCA3/MJD	SCA6	SCA7
Mean age at onset (years)	34	35	38	45	30
Onset after 55 years	-	-	±	++	-
Cerebellar syndrome	+++	+++	+++	+++	+++
Dysarthria	+++	+++	+++	++	+++
Babinski sign	++	+	++	0	++
Brisk reflexes	++	+	++	+	+++
Diminished or abolished reflexes	+	++	++	++	0
Spasticity in lower limbs	++	±	++	±	++
Amyotrophy	+	+	++	-	++
Extrapyramidal syndrome/dystonia	±	±	+	0	+
Myoclonus	-	++	±	-	
Nystagmus	++	+	+++	++	+
Ophthalmoplegia	++	++	++	0	++
Decreased saccade velocity	+	++	+	0	+++
Decreased visual acuity	0	0	0	0	+++
Bulging eyes	+	+	+	0	+
Myokymia	+	++	+	0	+
Decreased vibration sense	++	++	+	++	++
Dysphagia	++	++	++	++	++
Sphincter disturbances	++	++	++	++	++
Dementia	+	+	+	0	+
Tremor	-	+	±	±	+
Axonal neuropathy	++	+++	++	0	+
Decreased hearing acuity	0	0	0	0	+

Usually, normal and expanded alleles carry uninterrupted CAG repeats and there is no overlap between the normal and the pathological range. There are, however, two exceptions, *SCA1* and *SCA2*, in which most of the normal alleles are interrupted by 1 to 3 CAT and CAA, respectively, and can attain the size of small pathological expansions, albeit rarely.[61-63] Such interruptions probably stabilize the repeat sequence when present at these loci. In the *TBP* gene, both normal and expanded alleles carry CAA interruptions.

Anticipation

Instability is the molecular basis of a major feature of ADCA: the phenomenon of anticipation, i.e., the earlier onset and/or more severe course of the disease in successive generations. Due to the increase in the size of the expansion from generation to generation and to the negative correlation between expansion size and

Table 4. Recording of ocular movements in *SCAs*.[210] From reference 3 by courtesy of the Nature Publishing Group.

nystagmus	Saccade accuracy	Velocity	Gaze evoked
SCA1	Hypermetria	Normal	Absent
SCA2	Normal	Slow	Absent
SCA3	Hypometria	Normal	Present
SCA6	Hypometria	Normal	Present
SCA7	Normal	Slow	Present

the age at onset (Fig. 3), the mean age at onset of ADCAs decreases with successive generations. The greatest anticipation is therefore found in *SCA7* families, in which the expansion is very unstable. However, anticipation is usually overestimated because of observation biases[64] and only a few expansions, exceeding 100 repeat units, are associated with infantile or juvenile cases in *SCA2*[65] and *SCA7*.[53,59,66,67]

De novo *Mutations*

Because the disease is not transmitted by infantile or juvenile patients, anticipation should lead to extinction of the disorder in carrier families after a variable number of generations. In *SCA7*, in which there is marked anticipation, de novo mutations occur to replace the nontransmitted pathological alleles.[68] These neomutations resulted from the expansion of large normal alleles, often designated as intermediate alleles (IA), that contain from 28 to 35 CAG units (Fig. 4).

A study on Asian and Caucasian families showed that the relative frequency of *SCAs* is correlated with the frequency of intermediate alleles (IA) in a given population.[69] More evidence supporting the notion that IA represent a reservoir for de novo mutations comes from the study of polymorphisms within the *MJD1* and DRPLA genes.[37,70] As already confirmed in HD,[71] the degree of instability at the *SCA3* locus increases with the size of the normal alleles even before they reach the pathological range.[55]

Phenotypes and Neuropathology Characterizing Each Form

Identification of the mutations has led to detailed analysis of large series of genetically homogeneous patients. The major clinical features for *SCA1, SCA2, SCA3/ MJD, SCA6* and *SCA7* patients are given in Table 3.

Table 5. Neuropathological features associated with *SCA* mutations[90;120;168;211]

Structure	SCA1	SCA2	SCA3/MJD	SCA6	SCA7
Cerebral cortex	-	+	-	-	(+)
White matter	-	+	-	-	-
Globus pallidus	+external	+	++internal	-	+
Sub-thalamic nucleus	+	+	++	-	++
Substantia nigra	+	++	++	-	++
Pontine nuclei	+	+++	++	-	+
Inferior olives	+++	+++·	-	(+)	+++
Purkinje cells	+	++	(+)	+++	++
Dentate nucleus	++	-	++	(+)	++
Spinocerebellar tracts	++	-	+++	-	+
Corticospinal tracts	-	-	(+)		+
Anterior horn	+	+	+	-	+
Posterior column	+	+++	+	-	+

Spared; (+) slight alteration; + discrete; ++ moderate; +++ severe atrophy

AGE AT ONSET

First, in all sub-forms of *SCA*, age at onset varies widely among patients from the same family and is negatively correlated with CAG repeat size. The repeat length explains 50-80% of the variability in age at onset (Fig. 1), indicating that other factors influence pathogenesis.[72;73] Recently, the CAG repeat length of the retinoic-acid-induced 1 (*RAI1*) gene was identified as a potential modifier of age at disease onset for *SCA2*.[74] This result implicates *RAI1* as a possible contributor to *SCA2* neurodegeneration and raises the possibility that other CAG-containing proteins may play a role in the pathogenesis of other polyglutamine disorders.

In contrast to HD, homozygosity is reported to cause earlier onset in *SCA2*,[8] *SCA3*/MJD[75-77] and *SCA6*,[45,78,79] suggesting that allelic dosage partially influences clinical onset.

CLINICAL PRESENTATION IN PATIENTS

Decreased visual acuity leading to blindness, resulting from progressive macular dystrophy, characterizes the majority of *SCA7* patients. Cerebellar ataxia is usually the presenting sign in adults with onset after the age of 30 years and visual failure becomes manifest up to 45 years later.[53] In contrast, in juvenile or infantile cases, decreased visual acuity can manifest up to 10 years before ataxia.[80] This sign, may, however, be absent in *SCA7* patients with late onset but can also be confounded with age-related macular degeneration in elderly patients or present, although rarely, in other *SCAs*.[81]

Table 6. Influence of CAG repeat size on clinical features of $SCAs$[4]

Locus	CAG repeat expansion			
	Small	Medium	Large	Very large
	Late Onset ————————→Early onset ————————→Juvenile case			
SCA1		Cerebellar ataxia, increased reflexes	Amyotrophic lateral sclerosis-like	
SCA2	Postural tremor	Cerebellar ataxia,	Cerebellar ataxia, decreased reflexes	Fasciculations, myokymia, myoclonus, chorea, dementia dystonia, cardiac degeneration
failure, retinal				
SCA3	Axonal neuropathy, DOPA-responsive parkinsonism	Cerebellar ataxia, gaze-evoked nystagmus	Dystonia	
SCA6	Episodic ataxia	Pure cerebellar ataxia	Few associated signs after 10-years course	
SCA7	Cerebellar ataxia without visual loss	Cerebellar ataxia, macular degeneration	Visual loss before cerebellar syndrome	Cardiac failure

Table 7. Clinical symptoms influenced by disease duration.[53;63;64;90;109;110]

Mutation	SCA1	SCA2	SCA3/MJD	SCA7	SCA6
Signs	Dysphagia	Dysphagia	Dysphagia	Dysphagia	Dysarthria
	Sphincter	Sphincter	Sphincter	Sphincter	
	disturbances	disturbances	disturbances	disturbances	
	Amyotrophy	Mental			
	Hyporeflexia	Mental impairment			
	Deep sensory	Hyporeflexia			
	loss	Fasciculations			
	Supranuclear	Slow eye			
	gaze palsy	movements			

No other clinical sign is specifically associated with a given genotype (Table 3) because of the extreme variability in phenotype among families that is partially explained by CAG repeat length (see below).

Patients with the *SCA1* mutation usually have pyramidal signs with hyperreflexia and gait spasticity[82] associated with severe disease progression. An early decrease in saccade velocity and reduced tendon reflexes without extrapyramidal signs is suggestive of *SCA2*.[63,83-85] Cognitive changes, whether or not in the context of dementia are also a prominent feature of *SCA2* patients.[86;87] Both, *SCA3*/MJD and *SCA6* patients frequently present with cerebellar oculomotor signs, such as saccadic smooth pursuit, gaze-evoked nystagmus and diplopia. *SCA3*/MJD patients frequently have ophthalmoplegia or amyotrophy[88] that may be associated with extrapyramidal signs, myokymia and bulging eyes in patients with Portuguese ancestry.[89-92] *SCA6* patients, however, usually have later onset, slower disease progression and very few neurological signs in addition to cerebellar ataxia during the first decade,[64] a profile that closely resembles that of *SCA5* patients.[93,94] Episodic ataxia has been described as the presenting sign in some *SCA6* patients.[22] Patients with CAG expansion in the *TBP* gene (*SCA15*) may present with ADCA associated with dementia and extrapyramidal signs.[95]

Paraclinical investigations can also help to identify group differences. Increased motor conduction times in the central (>10 ms) and peripheral (>18 ms) nervous system are distinctive of the *SCA1* phenotype.[84] Recording of ocular movements might also be useful (Table 4), but there is some overlap of phenotypes.[96;97]

SCA patients present with varying degrees of cerebellar atrophy. *SCA3*/MJD is characterized by moderate cerebellar atrophy with severe pontine and spinal atrophy.[90] The vermis is generally more affected in *SCA2*, in which the pons can also be severely atrophied. Cerebral MRI of *SCA6* patients show pure and severe atrophy of the cerebellar vermis and hemispheres, whereas brainstem and cerebral hemispheres are spared.[44,64,98,99] Pontine atrophy, while rare, may occur in *SCA6* patients.[100] These features correlate well with the neuropathological observations.

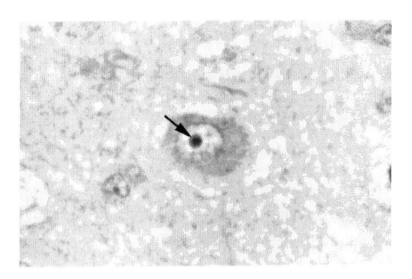

Figure 5. Intranuclear inclusions in the inferior olive of a *SCA7* patient carrying 85 CAG repeats (x250). The inclusion has been labeled with the 1C2 antibody[125] and revealed by the peroxidase/anti-peroxidase technique, with diaminobenzidine as the chromogen. Staining of the nucleus by Harris hematoxylin. These are also detected with an anti-ubiquitin antibody (data not shown). Reprinted with permission from reference 203 by courtesy of MARCEL DEKKER, Inc., and by reference 3 by courtesy of the Nature Publishing Group.

NEUROPATHOLOGICAL LESIONS

Each genetic sub-form has a strikingly different neuropathological profile (Table 5; for review, see ref. 101). *SCA1* patients have a widespread cell loss involving the spinal cord, vermis and basis ganglia.[90,102,103] The *SCA2* profile is considered typical of olivo-ponto-cerebellar atrophy since the inferior olive, substantia nigra, cerebellum (severe Purkinje cell loss) and pontine nuclei are affected.[104] It can be distinguished from *SCA1*, however, since the superior cerebellar peduncles are spared and the substantia nigra is severely lesioned. The cerebral cortex is also often affected. In *SCA3/MJD*, lesions of the basal ganglia (internal pallidum, sub-thalamic nucleus and substantia nigra), the intermediolateral column and Clarke's column are more severe than in *SCA1*, but the Purkinje cells, the inferior olives and posterior column are spared. This profile varies as a function of CAG repeat size.[90] *SCA6* patients have severe Purkinje cell loss with moderate degeneration of cells in the granular layer and inferior olives.

In *SCA7* patients, spinocerebellar, olivocerebellar and efferent cerebellar tracts are severely affected. Purkinje cells, granule cells and neurons in the dentate nuclei and the inferior olive, the substantia nigra and basis pontis also degenerate. The thalamus and striatum are spared. The distinctive features of *SCA7* are involvement

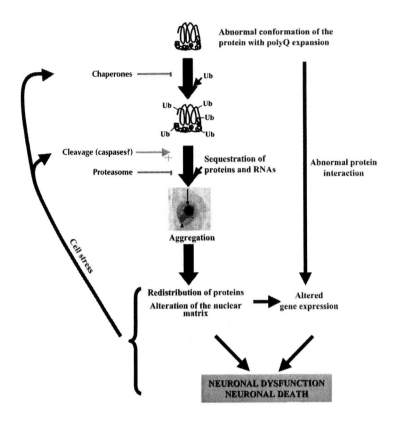

Figure 6. Putative mechanisms involved in polyQ diseases. The abnormal conformation of the protein, due to the polyQ expansion, leads to its ubiquitination and targeting to the proteasome that is unable, as are chaperones (except when over-expressed in models) to inhibit the aggregation of the protein with other components of the cell and finally leads to neuronal dysfunction and death. The aggregation process is accelerated by the cleavage of the protein. Alternatively, the alteration of the cellular functions could be due to abnormal protein-protein interactions.

of pregeniculate visual pathways and the retina. Pathological examination of the retina shows early degeneration of the photoreceptors, the bipolar and the granular cells, particularly in the foveal and parafoveal regions. Later, the inner retinal layers are affected with patchy loss of epithelial pigment cells and penetration of pigmented cells into the retinal layers.

FACTORS INFLUENCING CLINICAL VARIABILITY

The major factors that influence phenotype are the size of the repeat expansion and disease duration at examination.

CAG repeat size, is negatively correlated to age at onset (Fig. 3). It also has a major effect on phenotype expression. The rate of progression until death in SCA1,[105] SCA3/MJD[106] and SCA7[53] patients, is negatively correlated with repeat size. Indeed, large SCA7 expansions are associated with juvenile forms of the disease, which are more severe and progress more rapidly than adult forms.[53,66,67] The number of CAG repeats also affects the frequency of several clinical signs (Table 6) and partly accounts for phenotypic variability among patients.[53,63,64,90,107,108] Interestingly, cardiac failure was observed in patients with very large SCA2[65] and SCA7[67] expansions, indicating the possibility of extra-neurological involvement in extreme cases.

Part of the variability in phenotype can be explained by a bias resulting from clinical evaluation of patients with different disease durations. Frequency of particular signs increases with disease duration (Table 7).[53,63,64,90,109,110]

PHYSIOPATHOLOGY OF SPINOCEREBELLAR ATAXIAS CAUSED BY POLYGLUTAMINE EXPANSIONS

Expression of Ataxins

There is no homology among the proteins involved in these diseases except the polyglutamine tract and, in some instances, an adjacent polyproline-rich region. A significant homology is, however, found between ataxin-2 and the ataxin-2 related protein that does not include the polyglutamine tract.[6] Ataxin-7 shares a short and functional motif homologous to the phosphate binding site of arrestins, suggesting a phosphorylation-dependent binding of this protein to its partner(s) in the cell.[111] The presence of a core homologous to spliceosomal small nuclear ribonucleoproteins suggests the involvement of ataxin-2 in RNA splicing.[112]

The function of these proteins in the cell is not known, except in two cases: a sub-unit of a voltage-dependent gated channel (CACNA1A) responsible for SCA6 and the TATA binding protein involved in SCA15, a general transcription initiation factor that regulates the expression of most eukaryotic genes transcribed by RNA polymerase II.[12]

Apart from CACNA1A, which is preferentially expressed in Purkinje cells,[10] the site of major pathology in SCA6, the proteins are ubiquitously expressed in CNS and nonCNS tissues and there is no correlation between the distribution of the normal ataxins and the sites of the pathology.[113,114] Ataxin-1 is predominantly found in neuronal nuclei and the cytoplasm of peripheral tissues[115] and has RNA binding activity.[116] The protein ataxin-2 is exclusively cytoplasmic, with the strongest expression in Purkinje cells.[117] Ataxin-3 is a small cytoplasmic or nuclear protein.[118,119] Wild type ataxin-7 localizes in the cytoplasm of all the populations of neurons analyzed and colocalizes partially with BIP, a protein of the lumen of the endoplasmic reticulum, in control brains.[120] Nuclear labeling is observed in some brain neurons and cellular models, in which ataxin-7 interacts with the nuclear matrix.[121]

Mechanism of the Diseases

Taken together, several common features among the six forms of *SCAs* as well as of the other neurodegenerative diseases associated with CAG/polyQ expansion indicate a common toxic effect related to the expansion. They are all progressive and unremitting neurodegenerative diseases associated with a selective death of neurons in the CNS. The toxic property manifests above a given number of repeats, usually >35. Both normal and abnormal proteins are expressed at the same level in all tested tissues. Furthermore, there is no clear relationship between expression pattern and site of pathology, except for *SCA6*, expressed predominantly in Purkinje cells, and *SCA7*, where nuclear labeling of ataxin-7 was higher in structures with neuronal loss.[120]

Animal and cellular models have been very helpful for exploring the morphological and biochemical steps important for pathogenesis of CAG/polyQ repeat disorders. Directed expression of a human cDNA encoding the *SCA1*, *SCA2*, *SCA7* or truncated *SCA3* genes with expanded CAG repeats, and expression of an isolated expanded CAG repeat, caused degeneration and/or dysfunction of target cells in transgenic mice, mimicking the human disease.[122-124] Comparison of these models and those of other polyQ diseases with human pathology underlines several similarities and differences that allow a better understanding of the pathogenesis (Fig. 6).

A Gain of Function Leads to Aggregation

It is now well accepted that the disease is the result of a gain of function that occurs at the protein level, and increases with repeat size after a threshold of approximately 35-40 glutamines. This is in agreement with the expression of both the mutated and the normal proteins and the dominant nature of the mutation. The expansion probably alters the conformation of the polyQ tract as initially suggested by the specific detection of long repeats using the 1C2 antibody.[125,126] Similarly, ataxin-3 adopts a unique conformation that exposes the polyQ domain.[127] This could explain the formation of insoluble intranuclear or cytoplasmic aggregates detected in the brains of patients with these polyglutamine diseases as well as in animal[122] or cellular models.[128] These inclusions appear to constitute a common signature of this group of disorders (Fig. 5). In the case of *SCA2* and *SCA6*, they are cytoplasmic and nonubiquitinated.[129,130] A single study reported a nuclear aggregation in *SCA2* patients.[131] Alternatively, two groups reported the formation of cation channels mediated by the polyQ repeats.[132,133]

How are Inclusions Formed?

A number of hypotheses have been proposed, but none has been demonstrated in vivo: nonconvalent interactions with other proteins, transglutamination,[134,135] formation of multimeric aggregates by hydrogen-bonded polar zippers.[136] In vitro, the fibrillary appearance of inclusions under electron microscopy and the green birefringency after staining with Congo red, both of which are reminiscent of amyloid, are consistent with the polar zipper hypothesis.[137,138] The aggregation process depends on the concentration of the protein, the size of the polyQ repeat tract and increases with time and protein truncation.[139]

Polyglutamine expansions are also good substrates for transglutaminases in vitro,[140] and the presence of transglutaminase inhibitors prevented aggregate formation in COS-7 cells transiently transfected with truncated forms of the DRPLA or HD gene with expansions.[141,142]

Small tracts of polyalanine in the polyadenylation binding protein 2 can also aggregate in patients with oculo-pharyngeal-muscular dystrophy,[143] as modeled initially.[144] Indeed, it has been suggested that frameshift errors in the *MJD1* (*SCA3*) gene, which lead to alanine expansions, cause protein accumulation.[145]

Composition of Aggregates

Aggregates correspond to abnormal relocation of the protein with polyQ expansion. The corresponding normal proteins can also be recruited,[141,146,147] which is reminiscent of other inclusion diseases.[148]

The use of several antibodies directed against N- and C-terminal parts of the protein showed that only a truncated version, which includes the polyQ expansion, aggregates in the vast majority of these diseases, except for *SCA1* and *TBP*.[149] Caspase-mediated cleavage sites in ataxin-7 and ataxin-3 might be implicated as in huntingtin, atrophin and androgen receptor which are suspected to be truncated by such proteases[150] and, as a consequence, could have a toxic effect[151,152] and/or could more easily enter the nucleus[141,153] and/or more rapidly aggregate.[147,154] Evidence for aberrant proteolysis comes from experiments in which truncated proteins with polyglutamine expansions appeared more prone than full length proteins to aggregate or cause cell death by apoptosis.[146,147,151,155] In addition, truncated fragments, that may result from caspase-1 cleavage,[156] have been detected in nuclear aggregates of HD patients. The question remains whether this cleavage refers to the normal function of the proteins, to abnormal degradation or to a protective response of the cell to these proteins. Elucidation of the proteolytic processing mechanisms would help to clarify this point.

Ubiquitination of several aggregated proteins indicates their targeting to the proteasome, a major proteolytic system. Indeed, expanded ataxin-1, -3 and -7, huntingtin, atrophin and androgen receptor colocalise with the proteasome, and several studies have shown a redistribution of the proteasome complex to inclusions.[123,146,157-161]

Based on the hypothesis of an abnormal conformation of these proteins as a major event in these diseases, several groups investigated the involvement of chaperones and evidenced their expression and relocation in nuclear inclusions. There is also evidence from cellular and animal models of polyglutaminopathies that over-expression of heat-shock proteins HDJ2/HSDJ (HSP40 family) and HSP70 can slow down inclusion formation and cell death.[158,160-164]

Are Inclusions a Cause or a Consequence of the Pathogenic Process?

They are predominantly found in affected tissues and were detected before the phenotype in a mouse model of HD,[165,166] suggesting that they may be deleterious. Their presence, however, was not sufficient to initiate the degenerative process in epithelial cells of a *SCA3* Drosophila model with no phenotype or degeneration,[167] in unaffected tissues in *SCA7* patients[168] or in peripheral tissues in an HD mouse model.[169,170] Modified forms of ataxin-1 demonstrated that nuclear localization is a prerequisite for pathogenesis but that ubiquitination is not necessary,[171] as also shown in cellular models with HD constructs.[142] The inclusions may therefore only represent a pathological hallmark of the diseases and/or a cellular defense mechanism.[172] If the inclusions are not responsible for the initiation of the disease, they may be implicated in disease progression and severity.

Alteration of the Nuclear Functions

In *SCA1*, *SCA3*, *SCA7* and *SCA15*, the abnormal proteins relocalize to nuclear aggregates. In *SCA1*, this localization is essential for pathogenesis,[171] in contrast to ataxin-2 and *CACNA1A*, which remain cytoplasmic, at least in Purkinje cells.[130] Other studies have demonstrated that mutated ataxin-3, ataxin-7 and atrophin are associated with the nuclear matrix.[121,127,173] The presence of mutant ataxin-7 in the nucleoli might alter RNA synthesis and processing.[121] Similarly, the RNA binding activity of ataxin-1 diminishes with the increase in size of the polyglutamine tract of the protein.[116] Interestingly, mutant ataxin-1 and atrophin are responsible for a redistribution of the promyelocytic leukemia protein (PML) from the PML oncogenic domains (PODs) to the nucleoplasm and the inclusions, respectively.[174,175] These data suggest that an early event in the pathogenesis may be an alteration of the nuclear matrix. Furthermore, other components of the nuclear bodies (PODs) are sequestered to the inclusions, as are several transcriptional coactivators and corepressors (*TBP*, CBP, ETO/MTG8, P53, mSin3A,etc) and RNA.[173,174,176-180] In agreement with these observations, down-regulation of neuronal genes involved in signal transduction and calcium homeostasis precedes detectable pathology in *SCA1*[181,182] and HD mice[183,184] even in the absence of aggregation. Functional impairment is mirrored by morphological changes[130,183] or by biochemical abnormalities.[185]

Cell Death Mechanisms

Apoptosis is difficult to detect in brains of patients or animals affected by these late onset progressive diseases. However, a large amount of proteins usually involved in apoptosis are sequestered, redistributed or activated in polyQ disorders (caspases and components of the PODs). Caspases are activated[186-191] and can be recruited in inclusions.[190] Cytochrome c release is also observed in vitro.[191] Indeed, inhibition of caspases slowed down inclusion formation and cell death in HD and SCA1 models.[156,191] Transglutaminases are also effectors of apoptosis and were implicated in the aggregation process in several models.[141,142] Finally, in SCA3 and HD models, the c-Jun amino-terminal kinase is activated prior to cell death.[192,193]

Selectivity of Degeneration

The proteins with expanded polyglutamine tracts are widely expressed in the nervous system of patients, contrasting with the relatively selective pattern of degeneration observed in each disorder. There must, therefore, be cell-specific factors that help determine the selective vulnerability.

First, neurons are post-mitotic cells and cell-cycle arrest, by over-expression of p21/Waf1 in cellular models of SCA3, enhances the toxicity mediated by the polyQ expansion.[194] Indeed, inclusions in dividing cells disperse during mitosis, whereas they accumulate in differentiated cells such as neurons.[195]

Several studies have shown that the degree of somatic mosaicism detected in the nervous system does not account for the selectivity of neuronal death: i.e., the degree of mosaicism was lower in the cerebellum than in other brain regions.[56,196,197] A recent study in an HD mouse model, however, showed a mosaicism in striatal cells that was much greater than that previously observed in human tissues and that increased with age.[51]

Specificity might also result from specific interactions of the mutated protein with protein partners expressed preferentially in affected regions. Several protein partners have been identified, some of which have greater affinity for the expanded protein than the normal form. This is particularly the case with the leucine-rich acidic nuclear protein that colocalizes with SCA1 aggregates.[177]

A recent study showing a late degenerative process in Drosophila and mice models over-expressing the normal ataxin-1 points to a possible contribution of the expression level of the protein.[198]

Finally, selectivity is also a function of the progression of the disease that depends on CAG repeat size since infantile and adult cases can have different patterns of degeneration.

CONCLUSIONS

Discovery of the mutations underlying ADCA and the correlations between CAG repeat length and clinical or neuropathological features of genetically specific sub-forms of these diseases have simplified the molecular diagnosis and permit analysis of patients classified according to their genotype, a necessary step in the development of a precise nosology and a better follow up of the patients. Although it is often impossible to anticipate the *SCA* mutation on the basis of clinical criteria since phenotype depends on the locus, the size of the repeat expansion, the duration of the disease and other unknown factors, analysis of the molecular and clinical characteristics have revealed group differences that will be helpful for understanding the history and disease course of patients with ADCA.

The cause of the molecular instability and the pathophysiological consequences of the expanded polyQ tract remain partially unknown, and therapeutic intervention will require elucidation of the underlying pathogenic mechanism. Animal and cellular models are helping us to understand the processing of the pathological proteins and identify their molecular partners. New research areas are emerging and others are being refocused: level of protein expression,[198] dysfunction of the transcription machinery and mosaicism of the repeat in the CNS.[51] The identification of modifiers of disease severity, involved in RNA processing, transcriptional regulation and cellular detoxification, in an *SCA1 Drosophila* model is providing important clues to the pathological process and to the great variability in disease expression.[198]

Diagnosis Considerations

Identification of ADCA genes and their mutations enables routine diagnostic testing of individuals who already present with symptoms of the disease. Molecular analysis is also useful to distinguish disorders which are clinically similar or which may be confused with other diseases because of their extremely variable clinical presentation. DNA testing in asymptomatic at-risk individuals, however, raises many difficult ethical issues for severe adult-onset disorders for which no treatment can be proposed and for which age at onset cannot be precisely predicted from the number of CAG repeats. The international guidelines for Huntington's disease should also be followed for ADCA.[199]

Molecular diagnosis of isolated cases is also crucial. Positive isolated cases rarely carry de novo mutation.[68,200] More frequently, they reflect missing family histories if the transmitting parent died before onset of symptoms or is still asymptomatic because of marked anticipation.[63,84] In our experience, the yield of positive molecular testing in the absence of family history does not exceed 3%.

Case of *SCA6*

Although the mutational mechanism in *SCA6* is a translated, but small, CAG expansion, it is not clear whether the pathogenic mechanism is similar to the other dominant spinocerebellar ataxias caused by the same kind of mutation. An alteration of the calcium homeostasis has been implicated by several, albeit controversial, studies.[201-203]

TOWARDS THERAPY

There is no specific drug therapy for these neurodegenerative disorders. Currently, therapy remains purely symptomatic. Therapy in gain of function diseases with adult onset faces ethical and technical difficulties. Thanks to our knowledge of the function of *CACNA1A*, a temporary improvement could be obtained with acetazolamide in *SCA6* patients.[204] Several other major therapeutic avenues can be explored. Neurons could be protected by growth factors, such as CNTF[205] or BDNF.[142,206] Transplantation of cells to replace those that die produced encouraging results in *SCA1* mice[207] and, more recently, in HD patients.[208] The most promising strategy would still appear to be inhibition of the toxic effect of the polyQ expansion. This approach will, however, require better characterization of the pathological steps to allow early intervention in the pathogenic process and maintenance of cells in a functional state. Minocycline[209] that can delay disease progression in HD mice could be of benefit to patients with only a limited risk associated with the treatment.

ACKNOWLEDGMENTS

We apologize to authors whose publications could not be cited for reasons of space. The authors' works are financially supported by the VERUM foundation, the Institut National de la Santé et de la Recherche Médicale, the Association Française contre les Myopathies, the Ministère de la Recherche et de la Technologie (France) and the Association pour le Développement de la Recherche sur les Maladies Génétiques Neurologiques et Psychiatriques.

REFERENCES

1. Harding AE. Clinical features and classification of inherited ataxias. Adv Neurol 1993; 61:1-14.
2. Froment J, Bonnet P, Colrat A. Heredo-dégénérations rétinienne et spino-cérébelleuses: Variantes ophtalmoscopiques et neurologiques présentées par trois générations successives. J Med Lyon 1937; 22:153-163.
3. Stevanin G, Durr A, Brice A. Clinical and molecular advances in autosomal dominant cerebellar ataxias: from genotype to phenotype and physiopathology. Eur J Hum Genet 2000; 8:4-18.
4. Durr A, Brice A. Clinical and genetic aspects of spinocerebellar degeneration. Curr Opin Neurol 2000; 13:407-413.

5. Orr HT, Chung MY, Banfi S et al. Expansion of an unstable trinucleotide CAG repeat in spinocerebellar ataxia type 1. Nature Genet 1993; 4:221-226.

6. Pulst SM, Nechiporuk A, Nechiporuk T et al. Moderate expansion of a normally biallelic trinucleotide repeat in spinocerebellar ataxia type 2. Nature Genet 1996; 14:269-276.

7. Imbert G, Saudou F, Yvert G et al. Cloning of the gene for spinocerebellar ataxia 2 reveals a locus with high sensitivity to expanded CAG/glutamine repeats. Nature Genet 1996; 14:285-291.

8. Sanpei K, Takano H, Igarashi S et al. Identification of the spinocerebellar ataxia type 2 gene using a direct identification of repeat expansion and cloning technique, DIRECT. Nature Genet 1996; 14:277-284.

9. Kawaguchi Y, Okamoto T, Taniwaki M et al. CAG expansion in a novel gene for Machado-Joseph disease at chromosome 14q32.1. Nature Genet 1994; 8:221-227.

10. Zhuchenko O, Bailey J, Bonnen P et al. Autosomal dominant cerebellar ataxia (SCA6) associated with small polyglutamine expansions in the alpha $_{1A}$-voltage-dependent calcium channel. Nature Genet 1997; 15:62-69.

11. David G, Abbas N, Stevanin G et al. Cloning of the SCA7 gene reveals a highly unstable CAG repeat expansion. Nature Genet 1997; 17:65-70.

12. Koide R, Kobayashi S, Shimohata T et al. A neurological disease caused by an expanded CAG trinucleotide repeat in the TATA-binding protein gene: a new polyglutamine disease? Hum Mol Genet 1999; 8:2047-2053.

13. Koob MD, Moseley ML, Schut LJ et al. An untranslated CTG expansion causes a novel form of spinocerebellar ataxia (SCA8). Nature Genet 1999; 21:379-384.

14. Matsuura T, Yamagata T, Burgess DL et al. Large expansion of the ATTCT pentanucleotide repeat in spinocerebellar ataxia type 10. Nat Genet 2000; 26:191-194.

15. Holmes SE, O'Hearn EE, McInnis MG et al. Expansion of a novel CAG trinucleotide repeat in the 5' region of PPP2R2B is associated with SCA12. Nat Genet 1999; 23:391-392.

16. Vincent JB, Yuan QP, Schalling M et al. Long repeat tracts at SCA8 in major psychosis. Am J Med Genet 2000; 96:873-876.

17. Worth PF, Houlden H, Giunti P et al. Large, expanded repeats in SCA8 are not confined to patients with cerebellar ataxia. Nat Genet 2000; 24:214-215.

18. Stevanin G, Herman A, Durr A et al. Are (CTG)n expansions at the SCA8 locus rare polymorphisms? Nat Genet 2000; 24:213.

19. Giunti P, Stevanin G, Worth P et al. Molecular and clinical study of 18 families with ADCA type II: evidence for genetic heterogeneity and de novo mutation. Am J Hum Genet 1999; 64:1594-1603.

20. Devos D, Schraen-Maschke S, Vuillaume I et al. Clinical features and genetic analysis of a new form of spinocerebellar ataxia. Neurology 2001; 56:234-238.

21. Zoghbi HY, Orr HT. Glutamine repeats and neurodegeneration. Ann Rev Neurosci 2000; 23:217-247.

22. Jodice C, Mantuano E, Veneziano L et al. Episodic ataxia type 2 (EA2) and spinocerebellar ataxia type 6 (SCA6) due to CAG repeat expansion in the CACNA1A gene on chromosome 19p. Hum Mol Genet 1997; 6:1973-1978.

23. van Schaik IN, Jobsis GJ, Vermeulen M et al. Machado-Joseph disease presenting as severe asymmetric proximal neuropathy. J Neurol Neurosurg Psychiatry 1997; 63:534-536.

24. Pratt RTC. The Genetics of Neurological Disorders. London: Oxford University Press, 1967.

25. Hirayama K, Takayanagi T, Nakamura R et al. Spinocerebellar degenerations in Japan: A nationwide epidemiological and clinical study. Acta Neurol Scand 1994; 153 (Suppl.):1-22.

26. Leone M, Bottacchi E, D'Alessandro G et al. Hereditary ataxias and paraplegias in Valle d'Aosta, Italy: a study of prevalence and disability. Acta Neurol Scand 1995; 91:183-187.

27. Sridharan R, Radhakrishnan K, Ashok PP et al. Prevalence and pattern of spinocerebellar degenerations in northeastern Libya. Brain 1985; 108:831-483.

28. Orozco G, Estrada R, Perry TL et al. Dominantly inherited olivopontocerebellar atrophy from eastern Cuba. Clinical, neuropathological, and biochemical findings. J Neurol Sci 1989; 93:37-50.

29. Sequeiros J and Coutinho P. Epidemiology and clinical aspects of Machado-Joseph disease. Adv Neurol 1993; 61:139-153.

30. Wakisaka A, Sasaki H, Takada A et al. Spinocerebellar ataxia 1 (SCA1) in the Japanese in Hokkaido may derive from a single common ancestry. J Med Genet 1995; 32:590-592.

31. Didierjean O, Cancel G, Stevanin G et al. Linkage disequilibrium at the SCA2 locus. J Med Genet 1999; 36:415-417.

32. Saleem Q, Choudhry S, Mukerji M et al. Molecular analysis of autosomal dominant hereditary ataxias in the Indian population: high frequency of SCA2 and evidence for a common founder mutation. Hum Genet 2000; 106:179-187.

33. Stevanin G, Cancel G, Didierjean O et al. Linkage disequilibrium at the Machado-Joseph disease/Spinal cerebellar ataxia 3 locus: evidence for a common founder effect in French and Portuguese-Brazilian families as well as a second ancestral Portuguese-Azorean mutation. Am J Hum Genet 1995; 57:1247-1250.

34. Takiyama Y, Igarashi S, Rogaeva EA et al. Evidence for inter-generational instability in the CAG repeat in the MJD1 gene and for conserved haplotypes at flanking markers amongst Japanese and Caucasian subjects with Machado-Joseph disease. Hum Mol Genet 1995; 4:1137-1146.

35. Gaspar C, Lopes-Cendes I, DeStefano AL et al. Linkage disequilibrium analysis in Machado-Joseph disease patients of different ethnic origins. Hum Genet 1996; 98:620-624.

36. Endo K, Sasaki H, Wakisaka A et al. Strong linkage disequilibrium and haplotype analysis in Japanese pedigrees with Machado-Joseph disease. Am J Med Genet 1996; 67:437-444.

37. Stevanin G, Lebre AS, Mathieux C et al. Linkage disequilibrium between the spinocerebellar ataxia 3/Machado-Joseph disease mutation and two intragenic polymorphisms, one of which, X359Y, affects the stop codon. Am J Hum Genet 1997; 60:1548-1552.

38. Dichgans M, Schols L, Herzog J et al. Spinocerebellar ataxia type 6: Evidence for a strong founder effect among German families. Neurology 1999; 52:849-851.

39. Jonasson J, Juvonen V, Sistonen P et al. Evidence for a common Spinocerebellar ataxia type 7 (SCA7) founder mutation in Scandinavia. Eur J Hum Genet 2000; 8:918-922.

40. Stevanin G, David G, Durr A et al. Multiple origins of the spinocerebellar ataxia 7 (SCA7) mutation revealed by linkage disequilibrium studies with closely flanking markers, including an intragenic polymorphism (G3145TG/A3145TG). Eur J Hum Genet 1999; 7:889-896.

41. Mayo CD, Hernandez CJ, Cantarero DS et al. Distribution of dominant hereditary ataxias and Friedreich's ataxia in the Spanish population. Med Clin (Barc) 2000; 115:121-125.

42. Filla A, Mariotti C, Caruso G et al. Relative frequencies of CAG expansions in spinocerebellar ataxia and dentatorubropallidoluysian atrophy in 116 Italian families. Eur Neurol 2000; 44:31-36.

43. Silveira I, Coutinho P, Maciel P et al. Analysis of SCA1, DRPLA, MJD, SCA2, and SCA6 CAG repeats in 48 Portuguese ataxia families. Am J Med Genet 1998; 81:134-138.

44. Watanabe H, Tanaka F, Matsumoto M et al. Frequency analysis of autosomal dominant cerebellar ataxias in Japanese patients and clinical characterization of spinocerebellar ataxia type 6. Clin Genet 1998; 53:13-19.

45. Matsumura R, Futamura N, Fujimoto Y et al. Spinocerebellar ataxia type 6. Molecular and clinical features of 35 Japanese patients including one homozygous for the CAG repeat expansion. Neurology 1997; 49:1238-1243.

46. Pujana MA, Corral J, Gratacos M et al. Spinocerebellar ataxias in Spanish patients: genetic analysis of familial and sporadic cases. The Ataxia Study Group. Hum Genet 1999; 104:516-522.

47. Basu P, Chattopadhyay B, Gangopadhaya PK et al. Analysis of CAG repeats in SCA1, SCA2, SCA3, SCA6, SCA7 and DRPLA loci in spinocerebellar ataxia patients and distribution of CAG repeats at the SCA1, SCA2 and SCA6 loci in nine ethnic populations of eastern India. Hum Genet 2000; 106:597-604.

48. Gaspar C, Lopes-Cendes I, Hayes S et al. Ancestral origins of the Machado-Joseph disease mutation: A worldwide haplotype study. Am J Hum Genet 2001; 68:523-528.

49. Kang S, Jaworski A, Ohshima K et al. Expansion and deletion of CTG repeats from human disease genes are determined by the direction of replication in *E.coli*. Nature Genet 1995; 10:213-218.

50. Wells RD, Warren ST. Wells RD, Warren ST, eds. Genetic Instabilities and Hereditary Neurological Diseases. San Diego: Academic press, 1998.

51. Kennedy L and Shelbourne PF. Dramatic mutation instability in HD mouse striatum: does polyglutamine load contribute to cell-specific vulnerability in Huntington's disease? Hum Mol Genet 2000; 9:2539-2544.

52. Ranen NG, Stine OC, Abbott MH et al. Anticipation and instability of IT-15 (CAG)n repeats in parent- offspring pairs with Huntington disease. Am J Hum Genet 1995; 57:593-602.

53. David G, Dürr A, Stevanin G et al. Molecular and clinical correlations in autosomal dominant cerebellar ataxia with progressive macular dystrophy (*SCA7*). Hum Mol Genet 1998; 7:165-170.

54. Igarashi S, Takiyama Y, Cancel G et al. Intergenerational instability of the CAG repeat of the Machado-Joseph disease (MJD1) is affected by the genotype of the normal chromosome: Implications for the molecular mechanisms of the instability of the CAG repeat. Hum Mol Genet 1996; 5:923-932.

55. Takiyama Y, Sakoe K, Soutome M et al. Single sperm analysis of the CAG repeats in the gene for Machado- Joseph disease (MJD1): Evidence for nonMendelian transmission of the MJD1 gene and for the effect of the intragenic CGG/GGG polymorphism on theintergenerational instability. Hum Mol Genet 1997; 6:1063-1068.

56. Chong SS, McCall AE, Cota J et al. Gametic and somatic tissue-specific heterogeneity of the expanded SCA1 CAG repeat in spinocerebellar ataxia type 1. Nature Genet 1995; 10:344-350.

57. Cancel G, Abbas N, Stevanin G et al. Marked phenotypic heterogeneity associated with expansion of a CAG repeat sequence at the spinocerebellar ataxia 3/Machado-Joseph disease locus. Am J Hum Genet 1995; 57:809-816.

58. Monckton DG, Cayuela ML, Gould FK et al. Very large (CAG)(n) DNA repeat expansions in the sperm of two spinocerebellar ataxia type 7 males. Hum Mol Genet 1999; 8:2473-2478.

59. Gouw LG, Castaneda MA, McKenna CK et al. Analysis of the dynamic mutation in the SCA7 gene shows marked parental effects on CAG repeat transmission. Hum Mol Genet 1998; 7:525-532.

60. Grewal RP, Cancel G, Leeflang EP et al. French Machado-Joseph disease patients do not exhibit gametic segregation distortion: A sperm typing analysis. Hum Mol Genet 1999; 8:1779-1784.

61. Chung MY, Ranum LP, Duvick LA et al. Evidence for a mechanism predisposing to intergenerational CAG repeat instability in spinocerebellar ataxia type I. Nature Genet 1993; 5:254-258.

62. Quan F, Janas J, and Popovich BW. A novel CAG repeat configuration in the *SCA1* gene: Implications for the molecular diagnostics of spinocerebellar ataxia type 1. Hum Mol Genet 1995; 4:2411-2413.

63. Cancel G, Dürr A, Didierjean O et al. Molecular and clinical correlations in spinocerebellar ataxia 2: A study of 32 families. Hum Mol Genet 1997; 6:709-715.

64. Stevanin G, Durr A, David G et al. Clinical and molecular features of spinocerebellar ataxia type 6. Neurology 1997; 49:1243-1246.

65. Babovic-Vuksanovic D, Snow K, Patterson MC et al. Spinocerebellar ataxia type 2 (SCA 2) in an infant with extreme CAG repeat expansion. Am J Med Genet 1998; 79:383-387.

66. Johansson J, Forsgren L, Sandgren O et al. Expanded CAG repeat in Swedish Spinocerebellar ataxia type 7 (SCA7) patients: Effect of CAG repeat length on the clinical manifestation. Hum Mol Genet 1998; 7:171-176.

67. Benton CS, de Silva R, Rutledge SL et al. Molecular and clinical studies in SCA-7 define a broad clinical spectrum and the infantile phenotype. Neurology 1998; 51:1081-1086.

68. Stevanin G, Giunti P, Belal G et al. De novo expansion of intermediate alleles in spinocerebellar ataxia 7. Hum Mol Genet 1998; 7:1809-1813.

69. Takano H, Cancel G, Ikeuchi T et al. Close associations between prevalences of dominantly inherited spinocerebellar ataxias with CAG-repeat expansions and frequencies of large normal CAG alleles in Japanese and Caucasian populations. Am J Hum Genet 1998; 63:1060-1066.

70. Yanagisawa H, Fujii K, Nagafuchi S et al. A unique origin and multistep process for the generation of expanded DRPLA triplet repeats. Hum Mol Genet 1996; 5:373-379.

71. Goldberg YP, McMurray CT, Zeisler J et al. Increased instability of intermediate alleles in families with sporadic Huntington disease compared to similar sized intermediate alleles in the general population. Hum Mol Genet 1995; 4:1911-1918.

72. Ranum LP, Chung MY, Banfi S et al. Molecular and clinical correlations in spinocerebellar ataxia type I: evidence for familial effects on the age at onset. Am J Hum Genet 1994; 55:244-252.

73. DeStefano AL, Cupples LA, Maciel P et al. A familial factor independent of CAG repeat length influences age at onset of Machado-Joseph disease. Am J Hum Genet 1996; 59:119-127.

74. Hayes S, Turecki G, Brisebois K et al. CAG repeat length in RAI1 is associated with age at onset variability in spinocerebellar ataxia type 2 (SCA2). Hum Mol Genet 2000; 9:1753-1758.

75. Kawakami H, Maruyama H, Nakamura S et al. Unique features of the CAG repeats in Machado-Joseph disease. Nature Genet 1995; 9:344-345.

76. Lerer I, Merims D, Abeliovich D et al. Machado-Joseph disease: Correlation between the clinical features, the CAG repeat length and homozygosity for the mutation. Eur J Hum Genet 1996; 4:3-7.

77. Sobue G, Doyu M, Nakao N et al. Homozygosity for Machado-Joseph disease gene enhances phenotypic severity. J Neurol Neurosurg Psychiatry 1996; 60:354-356.

78. Geschwind DH, Perlman S, Figueroa KP et al. Spinocerebellar ataxia type 6. Frequency of the mutation and genotype- phenotype correlations. Neurology 1997; 49:1247-1251.

79. Ikeuchi T, Takano H, Koide R et al. Spinocerebellar ataxia type 6: CAG repeat expansion in alpha1A voltage- dependent calcium channel gene and clinical variations in Japanese population. Ann Neurol 1997; 42:879-884.

80. Abe T, Tsuda T, Yoshida M et al. Macular degeneration associated with aberrant expansion of trinucleotide repeat of the SCA7 gene in 2 Japanese families. Arch Ophthalmol 2000; 118:1415-1421.

81. Kouno R, Kawata A, Yoshida H et al. A family of SCA1 with pigmentary macular dystrophy. Rinsho Shinkeigaku 1999; 39:649-652.

82. Bürk K, Abele M, Fetter M et al. Autosomal dominant cerebellar ataxia type I: Clinical features and MRI in families with SCA1, SCA2 and SCA3. Brain 1996; 119:1497-1505.

83. Bürk K, Fetter M, Skalej M et al. Saccade velocity in idiopathic and autosomal dominant cerebellar ataxia. J Neurol Neurosurg Psychiatry 1997; 62:662-664.

84. Schols L, Amoiridis G, Buttner T et al. Autosomal dominant cerebellar ataxia: Phenotypic differences in genetically defined subtypes? Ann Neurol 1997; 42:924-932.

85. Wadia N, Pang J, Desai J et al. A clinicogenetic analysis of six Indian spinocerebellar ataxia (SCA2) pedigrees. The significance of slow saccades in diagnosis. Brain 1998; 121:2341-2355.

86. Burk K, Globas C, Bosch S et al. Cognitive deficits in spinocerebellar ataxia 2. Brain 1999; 122:769-777.

87. Filla A, De Michele G, Santoro L et al. Spinocerebellar ataxia type 2 in southern Italy: A clinical and molecular study of 30 families. J Neurol 1999; 246:467-471.

88. Matilla T, McCall A, Subramony SH et al. Molecular and clinical correlations in spinocerebellar ataxia type 3 and Machado-Joseph disease. Ann Neurol 1995; 38:68-72.

89. Stevanin G, Le Guern E, Ravise N et al. A third locus for autosomal dominant cerebellar ataxia type I maps to chromosome 14q24.3-qter: Evidence for the existence of a fourth locus. Am J Hum Genet 1994; 54:11-20.

90. Durr A, Stevanin G, Cancel G et al. Spinocerebellar ataxia 3 and Machado-Joseph disease: clinical, molecular and neuropathological features. Ann Neurol 1996; 39:490-499.

91. Schols L, Amoiridis G, Epplen JT et al. Relations between genotype and phenotype in German patients with the Machado-Joseph disease mutation. J Neurol Neurosurg Psychiatry 1996; 61:466-470.

92. Schols L, Vieira-Saecker AM, Schols S et al. Trinucleotide expansion within the MJD1 gene presents clinically as spinocerebellar ataxia and occurs most frequently in German SCA patients. Hum Mol Genet 1995;4:1001-1005.

93. Ranum LP, Schut LJ, Lundgren JK et al. Spinocerebellar ataxia type 5 in a family descended from the grandparents of President Lincoln maps to chromosome 11. Nature Genet 1994; 8:280-284.

94. Stevanin G, Herman A, Brice A et al. Clinical and MRI findings in spinocerebellar ataxia type 5. Neurology 1999; 53:1355-1357.

95. Nakamura K, Jeong SY, Ichikawa Y et al. SCA15, a novel autosomal dominant cerebellar ataxia caused by the expanded polyglutamine in TATA-binding protein identified with 1C2 antibody immunoscreening. Am J Hum Genet 2000; 67 (Suppl 2):2185.

96. Rivaud-Pechoux S, Durr A, Gaymard B et al. Eye movement abnormalities correlate with genotype in autosomal dominant cerebellar ataxia type I. Ann Neurol 1998; 43:297-302.

97. Burk K, Fetter M, Abele M et al. Autosomal dominant cerebellar ataxia type I: Oculomotor abnormalities in families with SCA1, SCA2, and SCA3. J Neurol 1999; 246:789-797.

98. Gomez CM, Thompson RM, Gammack JT et al. Spinocerebellar ataxia type 6: Gaze-evoked and vertical nystagmus, Purkinje cell degeneration, and variable age of onset. Ann Neurol 1997; 42:933-950.

99. Schols L, Kruger R, Amoiridis G et al. Spinocerebellar ataxia type 6: Genotype and phenotype in German kindreds. J Neurol Neurosurg Psychiatry 1998; 64:67-73.

100. Sugawara M, Toyoshima I, Wada C et al. Pontine atrophy in spinocerebellar ataxia type 6. Eur Neurol 2000; 43:17-22.

101. Iwabuchi K, Tsuchiya K, Uchihara T et al. Autosomal dominant spinocerebellar degenerations. Clinical, pathological, and genetic correlations. Rev Neurol (Paris) 1999; 155:255-270.

102. Robitaille Y, Schut L, and Kish SJ. Structural and immunocytochemical features of olivoponto-cerebellar atrophy caused by the spinocerebellar ataxia type 1 (SCA-1) mutation define a unique phenotype. Acta Neuropathol (Berl) 1995; 90:572-581.

103. Gilman S, Sima AA, Junck L et al. Spinocerebellar ataxia type 1 with multiple system degeneration and glial cytoplasmic inclusions. Ann Neurol 1996; 39:241-255.

104. Durr A, Smadja D, Cancel G et al. Autosomal dominant cerebellar ataxia type I in Martinique (French West Indies): Clinical and neuropathological analysis of 53 patients from three unrelated SCA2 families. Brain 1995; 118:1573-1581.

105. Jodice C, Malaspina P, Persichetti F et al. Effect of trinucleotide repeat length and parental sex on phenotypic variation in spinocerebellar ataxia I. Am J Hum Genet 1994; 54:959-965.

106. Klockgether T, Kramer B, Ludtke R et al. Repeat length and disease progression in spinocerebellar ataxia type 3. Lancet 1996; 348:830-830.

107. Tuite PJ, Rogaeva EA, St George-Hyslop PH et al. Dopa-responsive parkinsonism phenotype of Machado-Joseph disease: Confirmation of 14q CAG expansion. Ann Neurol 1995; 38:684-687.

108. Martin J, Van Regemorter N, Del-Favero J et al. Spinocerebellar ataxia type 7 (SCA7)—Correlations between phenotype and genotype in one large Belgian family. J Neurol Sci 1999; 168:37-46.

109. Durr A, Chneiweiss H, Khati C et al. Phenotypic variability in autosomal dominant cerebellar ataxia type I is unrelated to genetic heterogeneity. Brain 1993; 116:1497-1508.

110. Spadaro M, Giunti P, Lulli P et al. HLA-linked spinocerebellar ataxia: A clinical and genetic study of large Italian kindreds. Acta Neurol Scand 1992; 85:257-65.

111. Mushegian AR, Vishnivetskiy SA, Gurevich VV. Conserved phosphoprotein interaction motif is functionally interchangeable between ataxin-7 and arrestins. Biochemistry 2000; 39:6809-6813.

112. Neuwald AF, Koonin EV. Ataxin-2, global regulators of bacterial gene expression, and spliceosomal snRNP proteins share a conserved domain. J Mol Med 1998; 76:3-5.

113. Koyano S, Uchihara T, Fujigasaki H et al. Neuronal intranuclear inclusions in spinocerebellar ataxia type 2: Triple-labeling immunofluorescent study. Neurosci Lett 1999; 273:117-120.

114. Lindenberg KS, Yvert G, Muller K et al. Expression analysis of ataxin-7 mRNA and protein in human brain: evidence for a widespread distribution and focal protein accumulation. Brain Pathol 2000; 10:385-394.

115. Servadio A, Koshy B, Armstrong D et al. Expression analysis of the ataxin-1 protein in tissues from normal and spinocerebellar ataxia type 1 individuals. Nature Genet 1995; 10:94-98.

116. Yue S, Serra HG, Zoghbi HY et al. The spinocerebellar ataxia type 1 protein, ataxin-1, has RNA-binding activity that is inversely affected by the length of its polyglutamine tract. Hum Mol Genet 2001; 10:25-30.

117. Huynh DP, Del Bigio MR, Ho DH et al. Expression of ataxin-2 in brains from normal individuals and patients with Alzheimer's disease and spinocerebellar ataxia 2. Ann Neurol 1999; 45:232-241.

118. Paulson HL, Das SS, Crino PB et al. Machado-Joseph disease gene product is a cytoplasmic protein widely expressed in brain. Ann Neurol 1997; 41:453-462.

119. Schmidt T, Landwehrmeyer GB, Schmitt I et al. An isoform of ataxin-3 accumulates in the nucleus of neuronal cells in affected brain regions of SCA3 patients. Brain Pathol 1998; 8:669-679.

120. Cancel G, Duyckaerts C, Holmberg M et al. Distribution of ataxin-7 in normal human brain and retina. Brain 2000; 123:2519-2530.

121. Kaytor MD, Duvick LA, Skinner PJ et al. Nuclear localization of the spinocerebellar ataxia type 7 protein, ataxin-7. Hum Mol Genet 1999; 8:1657-1664.

122. Heintz N and Zoghbi HY. Insights from mouse models into the molecular basis of neurodegeneration. Annu Rev Physiol 2000; 62:779-802.

123. Yvert G, Lindenberg KS, Picaud S et al. Expanded polyglutamines induce neurodegeneration and trans-neuronal alterations in cerebellum and retina of SCA7 transgenic mice. Hum Mol Genet 2000; 9:2491-2506.

124. Lin CH, Tallaksen-Greene S, Chien WM et al. Neurological abnormalities in a knock-in mouse model of Huntington's disease. Hum Mol Genet 2001; 10:137-144.

125. Trottier Y, Lutz Y, Stevanin G et al. Polyglutamine expansion as a pathological epitope in Huntington's disease and four dominant cerebellar ataxias. Nature 1995; 378:403-406.

126. Stevanin G, Trottier Y, Cancel G et al. Screening for proteins with polyglutamine expansions in autosomal dominant cerebellar ataxias. Hum Mol Genet 1996; 5:1887-1892.

127. Perez MK, Paulson HL, and Pittman RN. Ataxin-3 with an altered conformation that exposes the polyglutamine domain is associated with the nuclear matrix. Hum Mol Genet 1999; 8:2377-2385.

128. Lunkes A and Mandel J-L. Polyglutamines, nuclear inclusions and neurodegeneration. Nature Med 1997; 3:1201-1202.

129. Ishikawa K, Fujigasaki H, Saegusa H et al. Abundant expression and cytoplasmic aggregations of [alpha]1A voltage- dependent calcium channel protein associated with neurodegeneration in spinocerebellar ataxia type 6. Hum Mol Genet 1999; 8:1185-1193.

130. Huynh DP, Figueroa K, Hoang N et al. Nuclear localization or inclusion body formation of ataxin-2 are not necessary for SCA2 pathogenesis in mouse or human. Nat Genet 2000; 26:44-50.

131. Koyano S, Uchihara T, Fujigasaki H et al. Neuronal intranuclear inclusions in spinocerebellar ataxia type 2. Ann Neurol 2000; 47:550.

132. Monoi H, Futaki S, Kugimiya S et al. Poly-L-glutamine forms cation channels: Relevance to the pathogenesis of the polyglutamine diseases. Biophys J 2000; 78:2892-2899.

133. Hirakura Y, Azimov R, Azimova R et al. Polyglutamine-induced ion channels: A possible mechanism for the neurotoxicity of Huntington and other CAG repeat diseases. J Neurosci Res 2000; 60:490-494.

134. Green H. Human genetic diseases due to codon reiteration: Relationship to an evolutionary mechanism. Cell 1993; 74:955-956.

135. Kahlem P, Terre C, Green H et al. Peptides containing glutamine repeats as substrates for transglutaminase-catalyzed cross-linking: Relevance to diseases of the nervous system. Proc Natl Acad Sci U S A 1996; 93:14580-14585.

136. Perutz MF, Johnson T, Suzuki M et al. Glutamine repeats as polar zippers: their possible role in inherited neurodegenerative diseases. Proc Natl Acad Sci USA 1994; 91:5355-5358.

137. Wanker EE. Protein aggregation and pathogenesis of Huntington's disease: Mechanisms and correlations. Biol Chem 2000; 381:937-942.

138. Hollenbach B, Scherzinger E, Schweiger K et al. Aggregation of truncated GST-HD exon 1 fusion proteins containing normal range and expanded glutamine repeats. Philos Trans R Soc Lond B Biol Sci 1999; 354:991-994.

139. Scherzinger E, Sittler A, Schweiger K et al. Self-assembly of polyglutamine-containing huntingtin fragments into amyloid-like fibrils: Implications for Huntington's disease pathology. Proc Natl Acad Sci U S A 1999; 96:4604-4609.

140. Kahlem P, Green H, and Djian P. Transglutaminase action imitates Huntington's disease: Selective polymerization of Huntingtin containing expanded polyglutamine. Mol Cell 1998; 1:595-601.

141. Igarashi S, Koide R, Shimohata T et al. Suppression of aggregate formation and apoptosis by transglutaminase inhibitors in cells expressing truncated DRPLA protein with an expanded polyglutamine stretch. Nature Genet 1998; 18:111-117.

142. Saudou F, Finkbeiner S, Devys D et al. Huntingtin acts in the nucleus to induce apoptosis but death does not correlate with the formation of intranuclear inclusions. Cell 1998; 95:55-66.

143. Brais B, Bouchard JP, Xie YG et al. Short GCG expansions in the PABP2 gene cause oculopharyngeal muscular dystrophy. Nature Genet 1998; 18:164-167.

144. Perutz MF. Glutamine repeats and inherited neurodegenerative diseases: molecular aspects. Curr Opin Struct Biol 1996; 6:848-858.

145. Gaspar C, Jannatipour M, Dion P et al. CAG tract of MJD-1 may be prone to frameshifts causing polyalanine accumulation. Hum Mol Genet 2000; 9:1957-1966.

146. Paulson HL, Perez MK, Trottier Y et al. Intranuclear inclusions of expanded polyglutamine protein in spinocerebellar ataxia type 3. Neuron 1997; 19:333-344.

147. Martindale D, Hackam A, Wieczorek A et al. Length of huntingtin and its polyglutamine tract influences localization and frequency of intracellular aggregates. Nature Genet 1998; 18:150-154.

148. Welch WJ, Gambetti P. Chaperoning brain diseases. Nature 1998; 392:23-24.

149. Wellington CL, Hayden MR. Caspases and neurodegeneration: On the cutting edge of new therapeutic approaches. Clin Genet 2000; 57:1-10.

150. Wellington CL, Ellerby LM, Hackam AS et al. Caspase cleavage of gene products associated with triplet expansion disorders generates truncated fragments containing the polyglutamine tract. J Biol Chem 1998; 273:9158-9167.

151. Ikeda H, Yamaguchi M, Sugai S et al. Expanded polyglutamine in the Machado-Joseph disease protein induces cell death in vitro and in vivo. Nature Genet 1996; 13:196-202.

152. Ellerby LM, Andrusiak RL, Wellington CL et al. Cleavage of atrophin-1 at caspase site aspartic acid 109 modulates cytotoxicity. J Biol Chem 1999; 274:8730-8736.
153. Hackam AS, Singaraja R, Zhang T et al. In vitro evidence for both the nucleus and cytoplasm as subcellular sites of pathogenesis in Huntington's disease. Hum Mol Genet 1999; 8:25-33.
154. Cooper JK, Schilling G, Peters MF et al. Truncated N-terminal fragments of huntingtin with expanded glutamine repeats form nuclear and cytoplasmic aggregates in cell culture. Hum Mol Genet 1998; 7:783-790.
155. Merry DE, Kobayashi Y, Bailey CK et al. Cleavage, aggregation and toxicity of the expanded androgen receptor in spinal and bulbar muscular atrophy. Hum Mol Genet 1998; 7:693-701.
156. Ona VO, Li M, Vonsattel JP et al. Inhibition of caspase-1 slows disease progression in a mouse model of Huntington's disease. Nature 1999; 399:263-267.
157. Bailey CK, McCampbell A, Madura K et al. Biochemical analysis of high molecular weight protein aggregates containing expanded polyglutamine repeat androgen receptor. Am J Hum Genet 1998; 63 (Suppl):A8.
158. Cummings CJ, Mancini MA, Antalffy B et al. Chaperone suppression of aggregation and altered subcellular proteasome localization imply protein misfolding in SCA1. Nature Genet 1998; 19:148-154.
159. Chai Y, Koppenhafer SL, Shoesmith SJ et al. Evidence for proteasome involvement in polyglutamine disease: Localization to nuclear inclusions in SCA3/MJD and suppression of polyglutamine aggregation in vitro. Hum Mol Genet 1999; 8:673-682.
160. Stenoien DL, Cummings CJ, Adams HP et al. Polyglutamine-expanded androgen receptors form aggregates that sequester heat shock proteins, proteasome components and SRC-1, and are suppressed by the HDJ-2 chaperone. Hum Mol Genet 1999; 8:731-741.
161. Wyttenbach A, Carmichael J, Swartz J et al. Effects of heat shock, heat shock protein 40 (HDJ-2), and proteasome inhibition on protein aggregation in cellular models of Huntington's disease. Proc Natl Acad Sci USA 2000; 97:2898-2903.
162. Chai Y, Koppenhafer SL, Bonini NM et al. Analysis of the role of heat shock protein (Hsp) molecular chaperones in polyglutamine disease. J Neurosci 1999; 19:10338-10347.
163. Warrick JM, Chan HY, Gray-Board GL et al. Suppression of polyglutamine-mediated neurodegeneration in Drosophila by the molecular chaperone HSP70. Nat Genet 1999; 23:425-428.
164. Muchowski PJ, Schaffar G, Sittler A et al. Hsp70 and hsp40 chaperones can inhibit self-assembly of polyglutamine proteins into amyloid-like fibrils. Proc Natl Acad Sci USA 2000; 97:7841-7846.
165. Mangiarini L, Sathasivam K, Seller M et al. Exon 1 of the HD gene with an expanded CAG repeat is sufficient to cause a progressive neurological phenotype in transgenic mice. Cell 1996; 87:493-506.
166. Davies SW, Turmaine M, Cozens BA et al. Formation of neuronal intranuclear inclusions (NII) underlies the neurological dysfunction in mice transgenic for the HD mutation. Cell 1997; 90:537-548.
167. Warrick JM, Paulson HL, Gray-Board GL et al. Expanded polyglutamine protein forms nuclear inclusions and causes neural degeneration in Drosophila. Cell 1998; 93:939-949.
168. Holmberg M, Duyckaerts C, Durr A et al. Spinocerebellar ataxia type 7 (SCA7): A neurodegenerative disorder with neuronal intranuclear inclusions. Hum Mol Genet 1998; 7:913-918.
169. Sathasivam K, Hobbs C, Turmaine M et al. Formation of polyglutamine inclusions in non-CNS tissue. Hum Mol Genet 1999; 8:813-822.
170. Fain JN, Del Mar NA, Meade CA et al. Abnormalities in the functioning of adipocytes from R6/2 mice that are transgenic for the Huntington's disease mutation. Hum Mol Genet 2001; 10:145-152.
171. Klement IA, Skinner PJ, Kaytor MD et al. Ataxin-1 nuclear localization and aggregation: Role in polyglutamine-induced disease in SCA1 transgenic mice. Cell 1998; 95:41-53.

172. Sisodia SS. Nuclear inclusions in glutamine repeat disorders: Are they pernicious, coincidental, or beneficial? Cell 1998; 95:1-4.
173. Wood JD, Nucifora FC, Duan K et al. Atrophin-1, the dentato-rubral and pallido-luysian atrophy gene product, interacts with ETO/MTG8 in the nuclear matrix and represses transcription. J Cell Biol 2000; 150:939-948.
174. Skinner PJ, Koshy BT, Cummings CJ et al. Ataxin-1 with an expanded glutamine tract alters nuclear matrix-associated structures. Nature 1997; 389:971-974.
175. Yamada M, Wood JD, Shimohata T et al. Widespread occurrence of intranuclear atrophin-1 accumulation in the central nervous system neurons of patients with dentatorubral-pallidoluysian atrophy. Ann Neurol 2001; 49:14-23.
176. Perez MK, Paulson HL, Pendse SJ et al. Recruitment and the role of nuclear localization in polyglutamine-mediated aggregation. J Cell Biol 1998; 143:1457-1470.
177. Matilla A, Koshy BT, Cummings CJ et al. The cerebellar leucine-rich acidic nuclear protein interacts with ataxin-1. Nature 1997; 389:974-978.
178. Steffan JS, Kazantsev A, Spasic-Boskovic O et al. The Huntington's disease protein interacts with p53 and CREB-binding protein and represses transcription. Proc Natl Acad Sci USA 2000; 97:6763-6768.
179. Shimohata T, Nakajima T, Yamada M et al. Expanded polyglutamine stretches interact with TAFII130, interfering with CREB-dependent transcription. Nat Genet 2000; 26:29-36.
180. McCampbell A, Taylor JP, Taye AA et al. CREB-binding protein sequestration by expanded polyglutamine. Hum Mol Genet 2000; 9:2197-2202.
181. Lin X, Antalffy B, Kang D et al. Polyglutamine expansion down-regulates specific neuronal genes before pathologic changes in SCA1. Nat Neurosci 2000; 3:157-163.
182. Vig PJ, Subramony SH, Qin Z et al. Relationship between ataxin-1 nuclear inclusions and Purkinje cell specific proteins in SCA-1 transgenic mice. J Neurol Sci 2000; 174:100-110.
183. Iannicola C, Moreno S, Oliverio S et al. Early alterations in gene expression and cell morphology in a mouse model of Huntington's disease. J Neurochem 2000; 75:830-839.
184. Li SH, Cheng AL, Li H et al. Cellular defects and altered gene expression in PC12 cells stably expressing mutant huntingtin. J Neurosci 1999; 19:5159-5172.
185. Tabrizi SJ, Cleeter MW, Xuereb J et al. Biochemical abnormalities and excitotoxicity in Huntington's disease brain. Ann Neurol 1999; 45:25-32.
186. Miyashita T, Matsui J, Ohtsuka Y et al. Expression of extended polyglutamine sequentially activates initiator and effector caspases. Biochem Biophys Res Commun 1999; 257:724-730.
187. Wang GH, Mitsui K, Kotliarova S et al. Caspase activation during apoptotic cell death induced by expanded polyglutamine in N2a cells. Neuroreport 1999; 10:2435-2438.
188. Moulder KL, Onodera O, Burke JR et al. Generation of neuronal intranuclear inclusions by polyglutamine-GFP: Analysis of inclusion clearance and toxicity as a function of polyglutamine length. J Neurosci 1999; 19:705-715.
189. Kouroku Y, Fujita E, Urase K et al. Caspases that are activated during generation of nuclear polyglutamine aggregates are necessary for DNA fragmentation but not sufficient for cell death. J Neurosci Res 2000; 62:547-556.
190. Sanchez I, Xu CJ, Juo P et al. Caspase-8 is required for cell death induced by expanded polyglutamine repeats. Neuron 1999; 22:623-633.
191. Li SH, Lam S, Cheng AL et al. Intranuclear huntingtin increases the expression of caspase-1 and induces apoptosis. Hum Mol Genet 2000; 9:2859-2867.
192. Yasuda S, Inoue K, Hirabayashi M et al. Triggering of neuronal cell death by accumulation of activated SEK1 on nuclear polyglutamine aggregations in PML bodies. Genes Cells 1999; 4:743-756.
193. Liu YF. Expression of polyglutamine-expanded huntingtin activates the SEK1-JNK pathway and induces apoptosis in a hippocampal neuronal cell line. J Biol Chem 1998; 273:28873-28877.

194. Yoshizawa T, Yamagishi Y, Koseki N et al. Cell cycle arrest enhances the in vitro cellular toxicity of the truncated Machado-Joseph disease gene product with an expanded polyglutamine stretch. Hum Mol Genet 2000; 9:69-78.
195. Rich T, Assier E, Skepper J et al. Disassembly of nuclear inclusions in the dividing cell—A novel insight into neurodegeneration. Hum Mol Genet 1999; 8:2451-2459.
196. Cancel G, Gourfinkel-An I, Stevanin G et al. Somatic mosaicism of the CAG repeat expansion in spinocerebellar ataxia type 3/Machado-Joseph disease. Human Mutation 1998; 11:23-27.
197. Matsuura T, Sasaki H, Yabe I et al. Mosaicism of unstable CAG repeats in the brain of spinocerebellar ataxia type 2. J Neurol 1999; 246:835-839.
198. Fernandez-Funez P, de Gouyon B et al. Identification of genes that modify ataxin-1-induced neurodegeneration. Nature 2000; 408:101-106.
199. World Federation of Neurology Research Group on Huntington's Chorea. International Huntington Association and the World Federation of Neurology Research Group on Huntington's Chorea. Guidelines for the molecular genetics predictive test in Huntington's disease. J Med Genet 1994; 31:555-559.
200. Schols L, Szymanski S, Peters S et al. Genetic background of apparently idiopathic sporadic cerebellar ataxia. Hum Genet 2000; 107:132-137.
201. Restituito S, Thompson RM, Eliet J et al. The polyglutamine expansion in spinocerebellar ataxia type 6 causes a beta subunit-specific enhanced activation of P/Q-type calcium channels in *Xenopus* oocytes. J Neurosci 2000; 20:6394-6403.
202. Matsuyama Z, Wakamori M, Mori Y et al. Direct alteration of the P/Q-type Ca2+ channel property by polyglutamine expansion in spinocerebellar ataxia 6. J Neurosci 1999;19:RC14.
203. Stevanin G, Durr A, Brice A. Spinocerebellar ataxia 7 (chapter 23). In: Klockgether T, ed. Handbook of ataxia disorders. NY: Marcel Dekker, 2001: 463-486.

SPINOCEREBELLAR ATAXIA TYPE 10: A DISEASE CAUSED BY A LARGE ATTCT REPEAT EXPANSION

Tohru Matsuura and Tetsuo Ashizawa

INTRODUCTION

Spinocerebellar ataxia type 10 (SCA10) is an autosomal dominant disease characterized by ataxia and seizures.[1-3] It belongs to a group of diseases known as autosomal dominant cerebellar ataxias (ADCAs). To date, 16 ADCA loci, including SCA1,[4] SCA2,[5-7] SCA3/Machado-Joseph disease (MJD),[8] SCA4,[9] SCA5,[10] SCA6,[11] SCA7,[12] SCA8,[13] SCA10,[2,3] SCA11,[14] SCA12,[15] SCA13,[16] SCA14,[17] SCA16,[17a] SCA17[17b,17c,17d] and dentatorubral-pallidoluysian atrophy (*DRPLA*),[18,19] have been mapped to specific chromosomal regions. While mutations involved in SCA4, SCA5, SCA11, SCA13, SCA14 and SCA16 have not been identified, six of these 14 ADCAs, including SCA1, SCA2, SCA3/MJD, SCA6, SCA7, SCA17 and DRPLA, have shown an expansion of a coding CAG trinucleotide repeat tract as the disease-causing mutation at the respective loci. In each of these diseases, the CAG repeat encodes a polyglutamine tract; therefore, an expansion of the CAG repeat gives rise to an elongation of the polyglutamine tract in the protein product. There is increasing evidence that the elongated polyglutamine tract leads to a gain of novel toxic function that causes the disease.[20-29] One exception may be SCA6, in which the polyglutamine tract is located in the alpha-1A calcium channel subunit (*CACNA1A*) gene, which shows a high level of expression in cerebellum.[11] The expansion size is 21-33 repeats, which would be within the normal range for other SCAs. Functional alterations of the P/Q type calcium channels in the cerebellum is a likely consequence although a gain of novel toxic function has also been postulated as the pathogenic mechanism of SCA6. In most ADCAs with polyglutamine expansions, the age of disease onset becomes progressively earlier in successive generations with

Department of Neurology, Baylor College of Medicine and Veterans Affairs Medical Center, Houston, Texas 77030 U.S.A.

increasing severity of the disorder;[6,7,18,19,30-35] this clinical phenomenon is known as anticipation. Studies on genotype-phenotype correlations showed that in these diseases anticipation is accompanied by an increasing size of expanded repeat in successive generations. Besides ADCAs, Huntington's disease (HD)[36] and Kennedy's disease[37] are also caused by expansions of polyglutamine-coding CAG repeats, and HD clearly shows anticipation.

It should be noted that not all ADCAs are caused by coded CAG repeat expansions. In SCA8 and SCA12, the trinucleotide repeat expansion mutations have been identified, but their disease-causing mechanisms differ from those with expanded polyglutamines. In SCA8, there is an expanded CTG repeat in the 3' untranslated region (UTR) of the *SCA8* gene that shows partial sequence complementation with *KLHL*-1, a gene on the opposite strand.[13] Since the *SCA8* does not show a detectable open reading frame, it has been postulated that the SCA8 transcript might function as a regulator of *KLHL-1* expression via its antisense activity.[38] However, the CTG repeat expansion at the SCA8 locus has been found in non-ataxia populations, and this has raised a controversy whether this repeat expansion is the disease-causing mutation. In SCA12, an CAG repeat is expanded in the 5' UTR of a protein phosphatase gene, *PPP2RB*.[15] The CAG repeat expansion increases the transcription of the *PPP2R2B* gene, which may contribute to the disease-causing mechanism.[15] The exact pathogenic roles of these repeat expansions need to be further investigated. Expanded trinucleotide repeat is also involved in the pathogenesis of an autosomal recessive neurological disease, Friedreich's ataxia, in which homozygous expansion of a GAA repeat located in the first intron of the *FRDA* gene causes the disease by decreasing the transcription of the *FRDA* gene.[39,40]

Neither clinical anticipation nor progressive expansions of the repeat in successive generations has been documented in SCA8, SCA12 and Friedreich's ataxia, although the expanded repeats are unstably transmitted in these diseases. Other diseases, such as myotonic dystrophy type 1 (DM1) caused by a CTG repeat expansion in the 3' UTR of the *DMPK* gene[41] and fragile X syndrome caused by a CGG repeat expansion in the 5' UTR of the *FMR1* gene,[42] show anticipation attributable to progressive expansions of the respective repeats in successive generations. In fragile X syndrome, the disease mechanism is a loss of function of the *FMR1* gene due to repressed transcription of *FMR1* with methylation. The pathogenic mechanism of DM1 appears to be complex; while gain of function mediated by the mutant DMPK mRNA in *trans* appears to play the major role,[43-45] loss of function of the genes in the vicinity, including *DMPK*,[46-49] *Six5*[50-53] and *DMWD*[54] may also have pathogenic importance.

In summary, the current data indicate that trinucleotide repeat expansions cause neurodegenerative disorders with a strong target predilection to the cerebellum. However, the pathogenic mechanism by which expanded repeats lead to the clinical phenotype varies, depending on the location of the repeat within the gene and sequence of the repeat unit. This chapter will review a novel type of repeat expansion disorder, SCA10, in which a large expansion of an intronic ATTCT pentanucleotide repeat has been identified as the disease-causing mutation.[56] The expanded ATTCT repeat allele at the SCA10 locus is unstable, and the disease exhibits autosomal dominant inheritance with anticipation. We will describe the clinical features, the strategies used for identifying the mutation, instability of the ATTCT repeat, genotype-phenotype correlation, population genetics, and potential pathogenic mechanisms.

CLINICAL FEATURES

Clinical characteristics of SCA10, which is now genetically defined by the ATTCT repeat expansion, is currently based on data obtained from six Mexican families.[1-3,55] The clinical phenotype of SCA10 is relatively homogeneous. The central feature of the clinical phenotype is cerebellar ataxia that usually starts as poor balance on gait. The gait ataxia gradually worsens with an increasing number of falls, necessitating use of a cane, a walker, and eventually a wheelchair. In an advanced stage, the patient becomes unable to stand or sit without support. Scanning dysarthria, which is a type of slurred speech typically seen in cerebellar ataxia, appears within a few years after the onset of gait ataxia. Scanning speech is due to ataxia involving the vocal cord, tongue, palate, cheek, and lip movements. Coordination of the diaphragm and other respiratory muscles are also impaired, contributing to the speech impairment. Poor coordination of tongue, throat, and mouth muscles also causes dysphagia in later stages of the disease. Dysphagia is not only a nuisance but often leads to life-threatening aspiration pneumonia. Severe dysphagia may require a percutaneous placement of a gastric tube for both prevention of aspiration and maintenance of nutritional intake. Hand coordination also starts deteriorating within a few years after the onset of gait ataxia. Handwriting and other fine motor tasks, such as buttoning cuffs, are first to be impaired, and followed by increasing difficulties in daily activities such as feeding, dressing, and personal hygiene. Tracking eye movements become abnormal, with fragmented pursuit, ocular dysmetria, and occasionally ocular flutter, which are all attributable to cerebellar dysfunction. Some patients with relatively severe ataxia show coarse gaze-induced nystagmus.

In addition to cerebellar ataxia, 20% to 60% of affected members of SCA10 families have recurrent seizures.[1-3,55] Most of these patients experience generalized motor seizures, but complex partial seizures have also been noted. An attack of complex partial seizure may occasionally be followed by a generalized motor seizure, suggesting secondary generalization of a focal seizure activity. In most cases, seizures are noted after the onset of gait ataxia. In untreated patients, generalized motor seizures could occur as frequent as daily and complex partial seizures may be even more frequent up to several times a day. However, conventional anticonvulsants such as phenytoin, carbamazepine, and valproic acid usually bring the seizures under reasonable control, although occasional breakthrough seizures may be noted. Seizure characteristics do not appear to change with age. However, seizure-related deaths have been noted in some affected members of SCA10 families (Grewal et al, personal communication). Interinctal electoencephalography shows evidence of cortical dysfunctions with or without focal epileptiform discharges in some patients.[54a]

While there is no overt progressive dementia, some SCA10 patients exhibit mild cognitive dysfunctions. Pyramidal and extrapyramidal dysfunction, visual impairment, hearing loss, peripheral neuropathy, and other nervous system abnormalities are usually absent, and if present, they are subtle. The combination of "pure" cerebellar ataxia and seizure is a phenotype unique to SCA10 and has not been seen in other ADCAs; patients with DRPLA who show this combination also have other conspicuous neurological abnormalities, which are rarely seen in SCA10. Studies of additional SCA10 families are necessary to further define the clinical phenotype.

Anticipation was first noted by Grewal et al[3] in their large SCA10 family. While anticipation is striking in this family, it was less prominent in another larger family described by Matsuura et al.[2] In small families, anticipation may be variable and difficult to evaluate.[54a] It is also noteworthy that severe early-onset phenotype has not been reported in SCA10, although cases with juvenile onset have been seen in some SCA10 families.

IDENTIFICATION OF THE *SCA10* MUTATION

Because SCA10 is an ADCA with anticipation[1-3] and several other SCA subtypes are associated with trinucleotide repeat expansions, the mutation responsible for SCA10 might also be an expansion of an unstable triplet repeat. Matsuura et al[2] and Zu et al[3] independently mapped the *SCA10* locus to the chromosome 22q13-qter region by linkage analyses. Two recombination events in these two families indicated that the *SCA10* gene resides within a 3.8-cM interval between *D22S1140* and *D22S1160*. Studies using additional polymorphic markers narrowed the SCA10 region to a 2.7-cM region between *D22S1140* and *D22S1153*.[55,56]

Chromosome 22 was the first human chromosome for which the Human Genome Project "completed" the sequencing.[57] However, while the entire euchromatic parts of chromosome 22 were sequenced, there were still 11 gaps that remain to be sequenced during the search of the SCA10 mutation. *D22S1160* and *D22S1153* resided in one of these gaps. Consequently, the exact physical size of the SCA10 candidate region was unknown. Nevertheless, two contigs composed of bacterial artificial chromosomes (BACs), phage P1-derived artificial chromosomes (PACs), and cosmids covered most of this region. The sequence data of these contigs enabled us to perform computer database searches for specific sequences in this region.

Meanwhile, additional four families with an autosomal dominant inheritance characterized by ataxia and seizures were identified. Although these families were relatively small to establish statistically significant linkage, the ataxia-seizure phenotype cosegreated with the SCA10 markers on chromosome 22. Because all these six families are of Mexican descent, their haplotypes of the SCA10 region were compared. The six families showed a common haplotype within the region (unpublished data), although the telomeric end of this region could not be defined due to a gap of the available contigs. In the SCA10 candidate interval in the chromosome 22

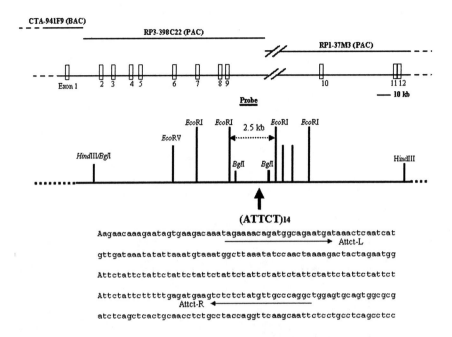

Figure 1. The physical map of the ATTCT pentanucleotide repeat region. *A*: A schematic presentation of the structure of the *E46* gene. *E46* consists of 12 exons. The ATTCT repeat is located in intron 9. The gap at the left of PAC 37M3 does not represent missing sequence, but was introduced to preserve scale. *B*: A restriction map of the ATTCT repeat region defined by flanking *Hind*III restriction sites (nt 17,023 and 34,567 by nucleotide positions in the PAC37M3 [GenBank accession # Z84478]). "Probe" indicates the position of the probe used (nt 25,222-26,021) to detect the 2.5 kb *EcoR*I fragment shown in Figure 4C in the Southern analysis. The ATTCT repeat is located downstream of the probe within the 2.5 kb *EcoR*I fragment. *C*: Nucleotide sequence of the ATTCT repeat (14 repeats; underlined, nt 26,101-26170) and the flanking regions (nt 25,981-26,281). Arrows underline PCR primer sequences (ATTCT-L and ATTCT-R) that were used for amplification of the ATTCT repeat region shown in Figure 5B. (From Nature Genetics 2000;21:191-194)

genome database at the Sanger Centre,[57] there were 14 trinucleotide repeats (>3 repeats in length) listed and they were screened for an expansion. However, none of them showed larger repeat size in SCA10 patients than in normal subjects. Moreover, repeat expansion detection (RED) analysis failed to show evidence of a CAG or CAA expansion.[56,58] Western blot analysis of proteins extracted from patients' lymphoblastoid cells using a monoclonal antibody raised against polyglutamine tracts also failed to detect abnormal proteins.[56,59] Because of the anticipation observed in SCA10, it was hypothesized that a non-triplet microsatellite repeat might be expanded in this disorder; hence a systematic search for various types of microsatellite sequences was initiated in this region.

By screening such repeats a pentanucleotide (ATTCT) repeat was found in intron 9 of the *E46L* gene (also known as *SCA10*) (Fig. 1).[56] Multiple tissue northern

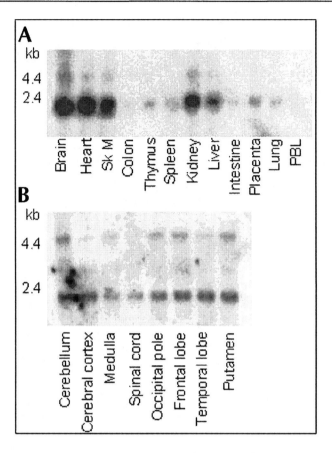

Figure 2. Northern blot analyses of multiple tissue blots (Clontech). A. Peripheral tissues. B. Central nervous system tissues. (From Nature Genetics 2000;21:191-194)

blots showed that the expression was widely noted throughout the brain, as well as in the skeletal muscle, heart, liver and kidney (Fig. 2).[56] The widespread expression of this gene throughout the brain was confirmed by an in situ hybridization study using mouse brain sections (Fig. 3).[56] PCR analysis showed repeat number polymorphisms in normal individuals. The repeat number ranged from 10 to 22 with 82.1 % heterozygosity in 604 chromosomes of three ethnic origins representing the Caucasian, Japanese and Mexican populations (Fig. 4).[56] Seqeunce analysis of the alleles obtained from 20 normal individuals showed tandem repeats of ATTCT without interruption. The allele distributions in each of the three ethnic populations were consistent with Hardy-Weinberg equilibrium.[56] In SCA10 families, PCR analysis demonstrated a uniform lack of heterozygosity of the ATTCT repeat alleles in all affected individuals and carriers of the disease haplotype, with the single allele of the ATTCT repeat shared by their unaffected parent. The single allele amplified

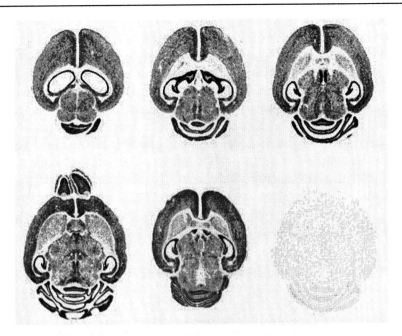

Figure 3. E46 is widely expressed in brain regions that are anatomical substrates for ataxia and epilepsy. E46 mRNA was detected by *in situ* hybridisation of radiolabeled probes to horizontal sections of 4-month-old adult (A-D) and 10 day old juvenile (C) mouse brain. Expression was similar to the pattern of cell density determined by cresyl violet staining of the same sections (not shown). A-D, dorsal to ventral progression; F, negative control for non-specific hybridisation to an adult brain section. (From Nature Genetics 2000;21:191-194)

from the affected parent is never transmitted to any of the affected offspring (Fig. 5).[56] Use of multiple PCR primer sets excluded the possibility that the lack of amplification is caused by a mutation within the sequence to which one of the PCR primers anneals. These data led us to postulate that the only the allele on the normal (non-SCA10) chromosome is amplified and perhaps the SCA10 chromosome has an expansion or other rearragements.

To investigate this hypothesis, southern blots of *Eco*RI fragments of the genomic DNA obtained from normal and SCA10 patients were analyzed with a probe (obtained by PCR amplification of DNA from a repeat-free region of PAC clone RP1-37M3), that corresponds to the region immediately upstream of the ATTCT repeat.[56] As predicted from the sequence data of the region, only a 2.5 kb fragment was detected in normal individuals. However, each affected family member showed a very large and variable size allele in addition to the expected normal 2.5 kb allele (Fig. 4). The variable size of expanded alleles among the patients and the absence of expanded alleles in over 600 normal chromosomes indicated that ATTCT repeat is expanded exclusively in SCA10 patients and unstable.

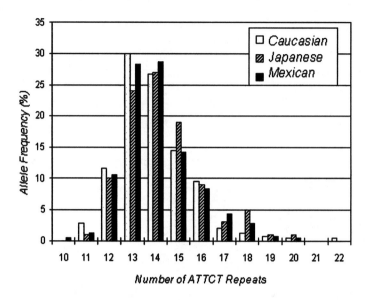

Figure 4. Distribution of the ATTCT repeat alleles in normal populations. Shown is a histogram of the normal ATTCT repeat alleles in Caucasian (n = 250), Japanese (n = 100) and Mexican (n = 254) chromosomes. (From Nature Genetics 2000;21:191-194)

PROSPECTS OF RESEARCH

Instability of the Expanded ATTCT Pentanucleotide Repeat

Extensive studies have been published on the instability of trinucleotide repeats, including exonic CAG repeats in SCA1,[60,61] SCA2,[62] SCA3/MJD,[63,64] SCA7,[34,35,65] DRPLA,[66-68] Kennedy's disease,[69] and Huntington's disease,[70-73] 3' UTR CTG repeats in DM1[74-77] and SCA8;[78] 5' UTR CGG/CCG repeats in Fragile X syndrome,[79] and FRAXE mental retardation;[80] 5' UTR CAG repeat in SCA12;[15] an intronic GAA repeat in Friedreich's ataxia;[81-83] and nonpathogenic CAG repeats in ERDA1[84] and SEF2.1.[85] CGG/CCG repeat expansions are also found at chromosomal fragile sites at FRAXF,[86] FRA11B,[87] and FRA16A,[88] and expansions at FRAXF and FRA11B are associated with mental retardation and Jacobsen syndrome, respectively. Although small pathogenic expansions of trinucleotide repeats in SCA6[11] and oculopharyngeal muscular dystrophy[89] show no or little repeat size instability, expanded alleles in most other diseases exhibit variable degrees of instability in somatic or germ line cells. Diseases that involve large repeat size expansions such as DM1, fragile X syndrome, FRAXE mental retardation and Friedreich's ataxia show greater degrees of instability. However, the size of expanded repeat is not the only determinant of the instability. The repeat unit sequence is clearly important, and different repeat units may be subjected to different mechanisms of

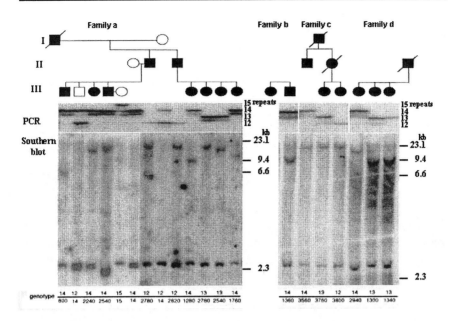

Figure 5. Expansion mutations in four SCA10 families. *a*: Pedigrees of the four families with ataxia and seizures studied for the SCA10 mutation. Square and round symbols indicate male and female members, respectively. Open symbols are asymptomatic individuals, and filled symbols indicate affected members. A diagonal line across a symbol denotes a deceased individual. *b*: PCR analysis of the ATTCT pentanucleotide repeat. All affected individuals showed a single allele of variable (note that each band accompanies a shadow band underneath due to PCR artifact). In Family a, two unaffected individuals (I-1 and III-2) are heterozygous and two spouses (II-1, III-5) are homozygous for the ATTCT repeat. In this family, affected individuals in the second generation (II-2 and II-3) failed to transmit their 12-repeat allele to their affected offspring (III-1, III-3, III-4, III-6, III-7, III-8 and III-9) while an unaffected offspring (III-2) received this allele from the affected father (II-2). The alleles of unaffected parents (I-1 and II-1) were passed on to their offspring in a pattern consistent with Mendelian inheritance. These data suggest that the affected individuals are apparently hemizygous for the ATTCT repeat. *c*: Southern analysis of expansion mutations of the ATTCT repeat region. Southern blots of the genomic DNA samples digested with *EcoR*I using a 0.8 % agarose gel show variably expanded alleles in affected members of the families shown above. All individual examined have a normal allele (2.5 kb). The apparent variability of the normal allele size is attributable to gel-loading artifacts since additional analyses using the same (*EcoR*I) and different (*EcoR*V, *Hind*III and *Bgl*I) restriction enzymes did not show consistent variability of the normal allele size. The genotype of each individual is shown at the bottom, with an estimated number of pentanucleotide repeats. (From Nature Genetics 2000;21:191-194)

instability from the view of DNA structure and stability. Among CAG/CTG expansion diseases, the CG content of the sequences surrounding the repeat tract has been correlated with the degree of repeat size instability per repeat unit.[90] DNA mismatch repair gene plays an important role in CAG repeat instability of HD transgenic mice.[91] Yet, there appear to be other genetic, epigenetic and environmental factors that may influence the repeat instabilities. Additional studies on minisatellite instability, including the dodecamer repeat involved in progressive myoclonus epilepsy type 1 (EPM1),[92-94] are also available in the literature.[95]

While these studies provide important insights about the instability of the ATTCT repeat, little is known about this novel class of disease-causing microsatellite repeat. Polymorphic pentanucleotide repeats have been reported in many human genes, including TTTTA repeats in *CYP11a*[96] and *apo(a)*,[97] CCTTT repeats in *NOS2*,[98] and (G/C)3NN repeats in *ETS-2* and dihydrofolate reductase genes,[99] among many others. Some pentanucleotide repeats are located in the 5' UTR or control region of genes, and may function as *cis*-acting elements that regulate the transcription of the downstream gene by affecting nucleosome assembly of the region.[99] However, none of these pentanucleotide repeats have been reported to show a pathogenic expansion. Expansions of the ATTCT repeat in SCA10 are among the largest of the microsatellite repeats involved in human diseases.[108] The location of the ATTCT repeat tract is also unusual; it is located in intron 9 of the *SCA10* gene. Further studies on the instability of this novel class of disease-causing microsatellite repeat are of unique scientific interest. Investigations of expanded ATTCT repeats for intergenerational changes, somatic and germ line instability, changes during development and aging, and instability in various experimental systems may provide important data for understanding the mechanism of this novel microsatellite instability.

Molecular Disease Mechanism of the ATTCT Expansion

The molecular mechanism by which the expanded ATTCT repeat causes the SCA10 phenotype remains to be investigated. The primary challenge is that *SCA10* is a novel gene of unknown function. *SCA10* consists of 12 exons spanning 172.8 kb, with the open reading frame of 1428 bp encoding 475 amino acids. Human *SCA10* is highly-conserved with its presumed mouse ortholog, *E46* (82% identity, 91% similarity over 475 amino acids). However, *SCA10* homologs of other species are largely unidentified; the next most similar sequence found in the GenBank database is a putative plant protein of unknown function identified by the *Arabidopsis* genome project (24% identical, 41% similar over 409 amino acids). Analysis of the amino acid sequence of the human SCA10 (E46L) protein suggests that it is a globular protein without transmembranous domains, nuclear localization signal or other type of signal peptide (Golgi, peroxisomal, vacuolar, or endoplasmic reticulum-retention). It does not appear to contain any known functional motifs, clusters or unusual patterns of charged amino acids or internal repeats of specific amino acid runs, and has an unremarkable predicted tertiary structure (data not shown).

Intron 9 of this gene is large (66,420 bp), raises the possibility that it might contain additional expressed sequences. An antisense transcript to *E46* could be disrupted by the pentanucleotide expansion and contribute to the SCA10 phenotype. To investigate this possibility, extensive sequence analysis of intron 9 was performed. The intron 9 region is rich for various repeat sequences and there are a total of 16 sequences with perfect identity to distinct ESTs in GenBank and one pseudogene apparently derived from CGI-47 gene on chromosome 3. Several lines of evidence strongly suggested these ESTs were likely to represent hnRNA or DNA contamination artifacts rather than functional transcribed sequences: (1) none were represented more than once, (2) none exhibited evidence of splicing relative to genomic DNA, (3) thirteen of the sixteen were oriented on the sense strand relative to E46, (4) several terminated at stretches of polyA nucleotide sequence, consistent with cryptic oligo-dT priming during cDNA synthesis, and (5) only 2 are derived from brain mRNA (both fetal). Finally, none of the EST sequences were positionally correlated with potential exons suggested by gene and exon identification programs including GRAIL, GENESCAN, FGENE, and HEXON. These analyses described above are equivalent to the NIX algorithms (http://www.hgmp.mrc.ac.uk/NIX).

At present, both loss-of-function (i.e., haploinsufficiency) and gain-of-function should be considered candidates for the pathogenic mechanism of the dominant inheritance in SCA10 (Fig 7). The ATTCT repeat is located in a large intron of the *SCA10* gene; the large expansion could therefore affect transcription or post-transcriptional processing of *SCA10*. Suppression of transcription by a large intronic repeat expansion has been documented in Friedreich's ataxia, where an expanded GAA repeat interferes with transcription of the *FRDA* gene.[100-102] Currently, the only available tissue from SCA10 patients are transformed lymphoblastoid cells, in which there were no alterations in the level of SCA10 mRNA by Northern blot analysis. However, one should be aware that lymphoblastoid cells do not show phenotype and the level of *SCA10* expression is substantially lower than other tissues such as brain, muscle, heart, liver and kidney. Thus, examination of the mRNA level in affected tissues of SCA10 patients is important. Investigating the functional role of *SCA10* is critical for understanding the pathogenic mechanism of SCA10, since *SCA10* is the prime candidate for the gene responsible for the disease. *SCA10* deficient mouse lines and cell lines would provide useful means to study the physiological functions of *SCA10*. We are also exploring proteins that interact with the SCA10 protein by yeast two-hybrid and immuno-co-precipitation technologies. Another possibility is that the expanded ATTCT repeat alters the splicing of the transcript. Aberrant splicing may give rise to a product with a gain of toxic function. Northern blot and RT-PCR analyses of mRNA in the lymphoblastoid cells from patients with SCA10 have not shown convincing abnormalities of the isoforms. Studies of the *SCA10* transcripts from the chromosome with an ATTCT repeat expansion may shed light on the pathophysiological mechanism of SCA10. Although this could be done using an exonic polymorphism within the *SCA10* gene, all available polymorphic markers have been homozygous in our SCA10 patients. Finding additional patients and informative polymorphic markers would facilitate the investigation. Other possible mechanisms include *cis* and *trans* effects of the ATTCT expansion on genes other than *SCA10*, resembling the putative pathophysiological mechanisms in DM1 in which an unstable CTG repeat expands up to several thousand copies in the 3' UTR of the *DMPK* gene.[103-107] However, the closest genes upstream and downstream are more than 200 kb away from the ATTCT repeat. The distance is far greater than the distances between the *DMPK* CTG repeat and adjacent genes, *DMWD* (~13 kb) and *SIX5* (~2 kb), although the *cis* effect of the ATTCT repeat expansion is difficult to predict.

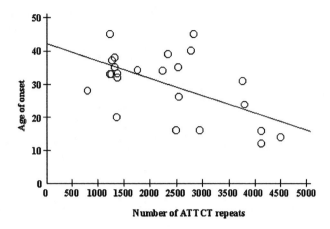

Figure 6. Correlation between the size of expanded SCA10 ATTCT repeat and the age of onset. A scatter plot shows an inverse correlation between the size of expansion and the age of onset in 26 SCA10 patients ($r^2 = 0.34$, $p = 0.018$). Each symbol represents an SCA10 patient, and the linear regression line is shown. (From Nature Genetics 2000;21:191-194)

Analogous to DM1, mRNA-mediated gain of function in *trans* is a viable possibility for the pathogenic mechanism. However, the repeat is located in intron 9, and it is unlikely that expanded AUUCU repeat is present in the mature SCA10 transcript. Our northern blot analysis of lymphoblastoid cells did not show expanded species of SCA10 mRNA. However, it is possible that the nuclear transcript with expanded AUUCU repeats may be processed differently in the nucleus. An expansion of an intronic CCTG tetranucleotide repeat has been found to be the pathogenic mutation of DM2, which gives rise to an accumulation of transcripts containing expanded CCUG repeats in nuclear foci similar to the foci containing expanded CUG repeats found in DM1.[108] Nuclear transcripts from expanded intronic repeats might exhibit similar gain-of-function mechanisms in these two diseases.

Genotype-Phenotype Correlation

Comparison of the clinical data and genotypes in our SCA10 patients revealed an inverse correlation between the expansion size and age of onset (Fig. 6). The number of the repeat ranged from 800 to 4500. Data on the genotype-phenotype correlation are currently limited to the inverse correlation between the size of the expanded ATTCT repeat and the age of onset.[56] However, this correlation is weak; indeed, some paternal transmissions have shown intergenerational contraction of the expanded repeat allele, in spite of the clinically observed anticipation.[56] A similar paradox in DM1 has been observed,[74,76,109] and it has been postulated that the apparent intergenerational contraction of the expanded CTG repeat is attributable to unusually strong somatic instability biased toward expansion in the father's leukocytes. Studies on the somatic and germ line instability of ATTCT repeats would

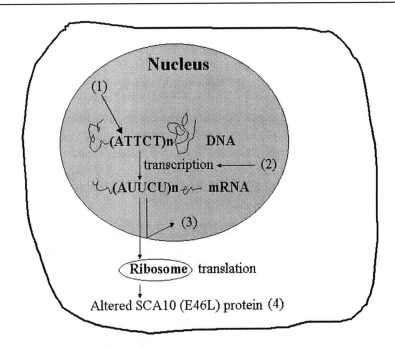

Figure7. Potential pathogenic mechanisms of SCA10. The pathogenic mechanism may involve both loss and gain of function at DNA, RNA and protein levels. (1) There might be proteins that specifically interact the ATTCT repeat, and the function of such proteins may be altered by an expansion of the ATTCT repeat. (2) An expansion of ATTCT repeat in intron 9 of the SCA10 gene might interfere with transcription of the gene. (3) Transcripts containing an expanded ATTCT repeat tract might be abnormally processed and possibly give rise to aberrant splicing, nuclear retention, or altered half life of the transcript. (4) Mechanisms described in (2) and/or (3) would quantitatively and/or qualitatively alter the protein product of the *SCA10* gene.

allow for investigation of a similar mechanism that could explain the observed paradox in SCA10.

Population Genetics

The exact prevalence of SCA10 is unknown. However, Rasmussen et al[110] examined a cohort of families from Mexico with inherited ataxia and found that SCA10 is the second most common inherited ataxia in Mexico after SCA2. All six of the SCA10 families identified to date are Mexican nationals or Mexican Americans. Whether SCA10 is unique to Mexicans or exists in other ethnic groups remains to be determined. However, the prevalence of SCA10 appears to be low in non-Mexican populations.[111,112] The origin of the SCA10 mutation is also of interest. The mutation could have arisen in the local Mexican population several generations ago, giving rise to a founder effect. Further haplotype analysis may provide an answer to this question. Alternatively, the mutation may have arisen in ethnic South American Indians or introduced into the current Mexican population by a Spanish conqueror. From a practical viewpoint, defining the at-risk population is an important issue for genetic counseling and may have interesting implications for population genetics.

ACKNOWLEDGMENTS

We thank the patients for cooperation. Work in the authors' laboratory was supported by grants from the Oxnard Foundation and National Ataxia Foundation (T.A.). We thank H.Y. Zoghbi, D.L. Nelson, D.L. Burgess, R.P. Grewal, J.F. Noebles, E. Alonso, and S.M. Pulst, for their collaboration and for useful suggestions and comments.

REFERENCES

1. Grewal RP, Tayag E, Figueroa KP et al. Clinical and genetic analysis of a distinct autosomal dominant spinocerebellar ataxia. Neurology. 1998; 51:1423-1426.
2. Matsuura T, Achari M, Khajavi M et al. Mapping of the gene for a novel spinocerebellar ataxia with pure cerebellar signs and epilepsy. Ann Neurol 1999; 45:407-411.
3. Zu L, Figueroa KP, Grewal R et al. Mapping of a new autosomal dominant spinocerebellar ataxia to chromosome 22. Am J Hum Genet 1999; 64:594-599.
4. Orr HT, Chung MY, Banfi S et al. Expansion of an unstable trinucleotide CAG repeat in spinocerebellar ataxia type 1. Nat Genet 1993; 4:221-226.
5. Pulst SM, Nechiporuk A, Nechiporuk T et al. Moderate expansion of a normally biallelic trinucleotide repeat in spinocerebellar ataxia type 2. Nat Genet 1996; 14:269-276.
6. Sanpei K, Takano H, Igarashi S et al. Identification of the spinocerebellar ataxia type 2 gene using a direct identification of repeat expansion and cloning technique, DIRECT. Nat Genet 1996; 14:277-284.
7. Imbert G, Saudou F, Yvert G et al. Cloning of the gene for spinocerebellar ataxia 2 reveals a locus with high sensitivity to expanded CAG/glutamine repeats. Nat Genet 1996; 14:285-291.
8. Kawaguchi Y, Okamoto T, Taniwaki M et al. CAG expansions in a novel gene for Machado-Joseph disease at chromosome 14q32.1. Nat Genet 1994; 8:221-228.
9. Flanigan K, Gardner K, Alderson K et al. Autosomal dominant spinocerebellar ataxia with sensory axonal neuropathy (SCA4): clinical description and genetic localization to chromosome 16q22.1. Am J Hum Genet 1996; 59:392-399.
10. Ranum LP, Schut LJ, Lundgren JK et al. Spinocerebellar ataxia type 5 in a family descended from the grandparents of President Lincoln maps to chromosome 11. Nat Genet 1994; 8:280-284.
11. Zhuchenko O, Bailey J, Bonnen P et al. Autosomal dominant cerebellar ataxia (SCA6) associated with small polyglutamine expansions in the alpha 1A-voltage-dependent calcium channel. Nat Genet 1997; 15:62-69.
12. David G, Abbas N, Stevanin G et al. Cloning of the SCA7 gene reveals a highly unstable CAG repeat expansion. Nat Genet 1997; 17:65-70.
13. Koob MD, Moseley ML, Schut LJ et al. An untranslated CTG expansion causes a novel form of spinocerebellar ataxia (SCA8). Nat Genet 1999; 21:379-84.
14. Worth PF, Giunti P, Gardner-Thorpe C et al. Autosomal dominant cerebellar ataxia type III: linkage in a large British family to a 7.6-cM region on chromosome 15q14-21.3. Am J Hum Genet 1999; 65:420-426.

15. Holmes SE, O'Hearn EE, McInnis MG et al. Expansion of a novel CAG trinucleotide repeat in the 5' region of PPP2R2B is associated with SCA12. Nat Genet 1999; 23:391-392.

16. Herman-Bert A, Stevanin G, Netter JC et al. Mapping of Spinocerebellar ataxia 13 to Chromosome 19q13.3-q13.4 in a Family with Autosomal Dominant Cerebellar Ataxia and Mental Retardation. Am J Hum Genet 2000; 67:229-235.

17. Yamashita I, Sasaki H, Yabe I et al. A novel locus for dominant cerebellar ataxia (SCA14) maps to a 10.2 cM interval flanked by D19S206 and D19S605 on chromosome 19q13.4-qter. Ann Neurol 2000; 48:156-163.

17a. Miyoshi Y, Yamada T, Tanimura M et al. A novel autosomal dominant spinocerebellar ataxia (SCA16) linked to chromosome 8q22.1-24.1. Neurology. 2001;57:96-100.

17b. Nakamura K, Jeong SY, Uchihara T et al. SCA17, a novel autosomal dominant cerebellar ataxia caused by an expanded polyglutamine in TATA-binding protein. Hum Mol Genet. 2001;10:1441-1448.

17c. Fujigasaki H, Martin JJ, De Deyn PP et al.CAG repeat expansion in the TATA box-binding protein gene causes autosomal dominant cerebellar ataxia. Brain. 2001;124:1939-1947.

17d. Zuhlke C, Hellenbroich Y, Dalski A et al. Different types of repeat expansion in the TATA-binding protein gene are associated with a new form of inherited ataxia. Eur J Hum Genet. 2001;9:160-164.

18. Koide R, Ikeuchi T, Onodera O et al. Unstable expansion of CAG repeat in hereditary dentatorubral-pallidoluysian atrophy (DRPLA). Nat Genet 1994; 6:9-13.

19. Nagafuchi S, Yanagisawa H, Sato K et al. Dentatorubral and pallidoluysian atrophy expansion of an unstable CAG trinucleotide on chromosome 12p. Nat Genet. 1994; 6:14-18.

20. Lin X, Antalffy B, Kang D et al. Polyglutamine expansion down-regulates specific neuronal genes before pathologic changes in SCA1. Nat Neurosci 2000; 3:157-163.

21. Schilling G, Wood JD, Duan K et al. Nuclear accumulation of truncated atrophin-1 fragments in a transgenic mouse model of DRPLA. Neuron 1999; 24:275-286.

22. Klement IA, Skinner PJ, Kaytor MD et al. Ataxin-1 nuclear localization and aggregation: role in polyglutamine-induced disease in SCA1 ransgenic mice. Cell 1998; 95:41-53.

23. Cummings CJ, Mancini MA, Antalffy B et al. Chaperone suppression of aggregation and altered subcellular proteasome localization imply protein misfolding in SCA1. Nat Genet 1998; 19:148-154.

24. Warrick JM, Paulson HL, Gray-Board GL et al. Expanded polyglutamine protein forms nuclear inclusions and causes neural degeneration in Drosophila. Cell 1998; 93:939-949.

25. Paulson HL, Perez MK, Trottier Y et al. Intranuclear inclusions of expanded polyglutamine protein in spinocerebellar ataxia type 3. Neuron 1997; 19:333-344.

26. Skinner PJ, Koshy BT, Cummings CJ et al. Ataxin-1 with an expanded glutamine tract alters nuclear matrix-associated structures. Nature 1997; 389:971-974.

27. Ikeda H, Yamaguchi M, Sugai S et al. Expanded polyglutamine in the Machado-Joseph disease protein induces cell death in vitro and in vivo. Nat Genet 1996; 13:196-202.

28. Holmberg M, Duyckaerts C, Durr A et al. Spinocerebellar ataxia type 7 (SCA7): a neurodegenerative disorder with neuronal intranuclear inclusions. Hum Mol Genet 1998; 7:913-918.

29. Koyano S, Uchihara T, Fujigasaki H et al. Neuronal intranuclear inclusions in spinocerebellar ataxia type 2: triple-labeling immunofluorescent study. Neurosci Lett 1999; 273:117-120.

30. Ranum LP, Chung MY, Banfi S et al. Molecular and clinical correlations in spinocerebellar ataxia type I: evidence for familial effects on the age at onset. Am J Hum Genet 1994; 55:244-252.

31. Maruyama H, Nakamura S, Matsuyama Z et al. Molecular features of the CAG repeats and clinical manifestation of Machado-Joseph disease. Hum Mol Genet 1995; 4:807-812.

32. Takiyama Y, Igarashi S, Rogaeva EA et al. Evidence for inter-generational instability in the CAG repeat in the MJD1 gene and for conserved haplotypes at flanking markers amongst Japanese and Caucasian subjects with Machado-Joseph disease. Hum Mol Genet 1995; 4:1137-1146.

33. Benton CS, de Silva R, Rutledge SL et al. Molecular and clinical studies in SCA-7 define a broad clinical spectrum and the infantile phenotype. Neurology 1998; 51:1081-1086.
34. David G, Durr A, Stevanin G et al. Molecular and clinical correlations in autosomal dominant cerebellar ataxia with progressive macular dystrophy (SCA7). Hum Mol Genet 1998; 7:165-170.
35. Gouw LG, Castaneda MA, McKenna CK et al. Analysis of the dynamic mutation in the SCA7 gene shows marked parental effects on CAG repeat transmission. Hum Mol Genet 1998; 7:525-532.
36. The Huntington's Disease Collaborative Research Group. A novel gene containing a trinucleotide repeat that is expanded and unstable on Huntington's disease chromosomes. Cell 1993; 72:971-983.
37. La Spada AR, Wilson EM, Lubahn DB et al. Androgen receptor gene mutations in X-linked spinal and bulbar muscular atrophy. Nature 1991; 352:77-79.
38. Nemes JP, Benzow KA, Koob MD. The SCA8 transcript is an antisense RNA to a brain-specific transcript encoding a novel actin-binding protein (KLHL1). Hum Mol Genet 2000; 9:1543-1551.
39. Bidichandani SI, Ashizawa T, Patel PI. The GAA triplet-repeat expansion in Friedreich ataxia interferes with transcription and may be associated with an unusual DNA structure. Am J Hum Genet 1998; 62:111-121.
40. Campuzano V, Montermini L, Molto MD et al. Friedreich's ataxia: autosomal recessive disease caused by an intronic GAA triplet repeat expansion. Science 1996; 271:1423-1427.
41. Fu YH, Pizzuti A, Fenwick RG Jr et al. An unstable triplet repeat in a gene related to myotonic muscular dystrophy. Science 1992; 255:1256-1258.
42. Fu YH, Kuhl DP, Pizzuti A et al. Variation of the CGG repeat at the fragile X site results in genetic instability: resolution of the Sherman paradox. Cell 1991; 67:1047-1058.
43. Timchenko LT, Miller JW, Timchenko NA et al. Identification of a (CUG)n triplet repeat RNA-binding protein and its expression in myotonic dystrophy. Nucleic Acids Res 1996; 24:4407-4414.
44. Roberts R, Timchenko NA, Miller JW et al. Altered phosphorylation and intracellular distribution of a (CUG)n triplet repeat RNA-binding protein in patients with myotonic dystrophy and in myotonin protein kinase knockout mice. Proc Natl Acad Sci USA 1997; 94:13221-13226.
45. Philips AV, Timchenko LT, Cooper TA. Disruption of splicing regulated by a CUG-binding protein in myotonic dystrophy. Science 1998; 280:737-741.
46. Mounsey JP, Mistry DJ, Ai CW et al. Skeletal muscle sodium channel gating in mice deficient in myotonic dystrophy protein kinase. Hum Mol Genet 2000; 9:2313-2320.
47. Berul CI, Maguire CT, Gehrmann J et al. Progressive atrioventricular conduction block in a mouse myotonic dystrophy model J Interv Card Electrophysiol 2000; 4:351-358.
48. Krahe R, Ashizawa T, Abbruzzese C et al. Effect of myotonic dystrophy trinucleotide repeat expansion on DMPK transcription and processing. Genomics 1995; 28:1-14.
49. Wang J, Pegoraro E, Menegazzo E et al. Myotonic dystrophy: evidence for a possible dominant-negative RNA mutation. Hum Mol Genet 1995; 4:599-606.
50. Sarkar PS, Appukuttan B, Han J et al. Heterozygous loss of Six5 in mice is sufficient to cause ocular cataracts. Nat Genet 2000; 25:110-114.
51. Klesert TR, Cho DH, Clark JI et al. Mice deficient in Six5 develop cataracts: implications for myotonic dystrophy. Nat Genet 2000; 25:105-109.
52. Thornton CA, Wymer JP, Simmons Z et al. Expansion of the myotonic dystrophy CTG repeat reduces expression of the flanking DMAHP gene. Nat Genet 1997; 16:407-409.
53. Klesert TR, Otten AD, Bird TD et al. Trinucleotide repeat expansion at the myotonic dystrophy locus reduces expression of DMAHP. Nat Genet 1997; 16:402-406.
54. Alwazzan M, Newman E, Hamshere MG et al. Myotonic dystrophy is associated with a reduced level of RNA from the DMWD allele adjacent to the expanded repeat. Hum Mol Genet. 1999; 8:1491-1497.

54a. Rasmussen A, Matsuura T, Ruano L et al. Clinical and genetic analysis of four Mexican families with spinocerebellar ataxia type 10. Ann Neurol 2001; 50:234-239.

55. Matsuura T, Watase K, Nagamitsu S et al. Fine mapping of the spinocerebellar ataxia type 10 region and search for a polyglutamine expansion. Ann Neurol 1999; 46:480.

56. Matsuura T, Yamagata T, Burgess DL et al. Large Expansion of the ATTCT pentanucleotide repeat in spinocerebellar ataxia type 10. Nat Genet 2000; 26:191-194.

57. Dunham I, Shimizu N, Roe BA et al. The DNA sequence of human chromosome 22. Nature 1999; 402:489-495.

58. Schalling M, Hudson TJ, Buetow KH et al. Direct detection of novel expanded trinucleotide repeats in the human genome. Nat Genet 1993; 4:135-139.

59. Trottier Y, Lutz Y, Stevanin G et al. Polyglutamine expansion as a pathological epitope in Huntington's disease and four dominant cerebellar ataxias. Nature 1995; 378:403-406.

60. Chung MY, Ranum LP, Duvick LA et al. Evidence for a mechanism predisposing to intergenerational CAG repeat instability in spinocerebellar ataxia type I. Nat Genet 1993; 5:254-258.

61. Chong SS, McCall AE, Cota J et al. Gametic and somatic tissue-specific heterogeneity of the expanded SCA1 CAG repeat in spinocerebellar ataxia type 1. Nat Genet 1995; 10:344-350.

62. Cancel G, Durr A, Didierjean O et al. Molecular and clinical correlations in spinocerebellar ataxia 2: a study of 32 families. Hum Mol Genet 1997; 6:709-715.

63. Cancel G, Abbas N, Stevanin G et al. Marked phenotypic heterogeneity associated with expansion of a CAG repeat sequence at the spinocerebellar ataxia 3/Machado-Joseph disease locus. Am J Hum Genet 1995; 57:809-816.

64. Cancel G, Gourfinkel-An I, Stevanin G et al. Somatic mosaicism of the CAG repeat expansion in spinocerebellar ataxia type 3/Machado-Joseph disease. Hum Mutat 1998; 11:23-27.

65. Monckton DG, Cayuela ML, Gould FK et al. Very large (CAG)$_n$ DNA repeat expansions in the sperm of two spinocerebellar ataxia type 7 males. Hum Mol Genet 1999; 8:2473-2478.

66. Ueno S, Kondoh K, Kotani Y et al. Somatic mosaicism of CAG repeat in dentatorubral-pallidoluysian atrophy (DRPLA). Hum Mol Genet 1995; 4:663-666.

67. Takano H, Onodera O, Takahashi H et al. Somatic mosaicism of expanded CAG repeats in brains of patients with dentatorubral-pallidoluysian atrophy: cellular population-dependent dynamics of mitotic instability. Am J Hum Genet 1996; 58:1212-1222.

68. Takiyama Y, Sakoe K, Amaike M et al. Single sperm analysis of the CAG repeats in the gene for dentatorubral-pallidoluysian atrophy (DRPLA): the instability of the CAG repeats in the DRPLA gene is prominent among the CAG repeat diseases. Hum Mol Genet 1999; 8:453-457.

69. Tanaka F, Reeves MF, Ito Y et al. Tissue-Specific Somatic Mosaicism in Spinal and Bulbar Muscular Atrophy Is Dependent on CAG-Repeat Length and Androgen receptor-Gene Expression Level. Am J Hum Genet 1999; 65:966-973.

70. Telenius H, Kremer B, Goldberg YP et al. Somatic and gonadal mosaicism of the Huntington disease gene CAG repeat in brain and sperm. Nat Genet 1994; 6:409-414.

71. Telenius H, Almqvist E, Kremer B et al. Somatic mosaicism in sperm is associated with intergenerational (CAG)n changes in Huntington disease. Hum Mol Genet 1995; 4:189-195.

72. Leeflang EP, Zhang L, Tavare S et al. Single sperm analysis of the trinucleotide repeats in the Huntington's disease gene: quantification of the mutation frequency spectrum. Hum Mol Genet 1995; 4:1519-1526.

73. Leeflang EP, Tavare S, Marjoram P et al. Analysis of germline mutation spectra at the Huntington's disease locus supports a mitotic mutation mechanism. Hum Mol Genet 1999; 8:173-183.

74. Ashizawa T, Anvret M, Baiget M et al. Characteristics of intergenerational contractions of the CTG repeat in myotonic dystrophy. Am J Hum Genet 1994; 54:414-423.

75. Ashizawa T, Monckton DG, Vaishnav S et al. Instability of the expanded (CTG)n repeats in the myotonin protein kinase gene in cultured lymphoblastoid cell lines from patients with myotonic dystrophy. Genomics 1996; 36:47-53.

76. Monckton DG, Wong LJ, Ashizawa T et al. Somatic mosaicism, germline expansions, germline reversions and intergenerational reductions in myotonic dystrophy males: small pool PCR analyses. Hum Mol Genet 1995; 4:1-8.

77. Wong LJ, Ashizawa T, Monckton DG et al. Somatic heterogeneity of the CTG repeat in myotonic dystrophy is age and size dependent. Am J Hum Genet 1995; 56:114-122.

78. Moseley ML, Schut LJ, Bird TD et al. SCA8 CTG repeat: en masse contractions in sperm and intergenerational sequence changes may play a role in reduced penetrance. Hum Mol Genet 2000; 9:2125-2130.

79. Eichler EE, Holden JJA, Popovich BW et al. Length of uninterrupted CGG repeats determines instability in the FMR1 gene. Nat Genet 1994; 8:88-94.

80. Knight SJL, Voelckel, MA, Hirst MC et al. Triplet expansion at the FRAXE locus and X-linked mild mental handicap. Am J Hum Genet 1994; 55:81-86.

81. Montermini L, Kish SJ, Jiralerspong S et al. Somatic mosaicism for Friedreich's ataxia GAA triplet repeat expansions in the central nervous system. Neurology 1997; 49:606-610.

82. De Michele G, Cavalcanti F, Criscuolo C et al. Parental gender, age at birth and expansion length influence GAA repeat intergenerational instability in the X25 gene: pedigree studies and analysis of sperm from patients with Friedreich's ataxia. Hum Mol Genet 1998; 7:1901-1906.

83. Bidichandani SI, Purandare SM, Taylor EE et al. Somatic sequence variation at the Friedreich ataxia locus includes complete contraction of the expanded GAA triplet repeat, significant length variation in serially passaged lymphoblasts and enhanced mutagenesis in the flanking sequence. Hum Mol Genet 1999; 8:2425-2436.

84. Ikeuchi T, Sanpei K, Takano H et al. A novel long and unstable CAG/CTG trinucleotide repeat on chromosome 17q. Genomics 1998; 49:321-326.

85. Breschel TS, McInnis MG, Margolis RL et al. A novel, heritable, expanding CTG repeat in an intron of the SEF2-1 gene on chromosome 18q21.1. Hum Mol Genet 1997; 6:1855-1863.

86. Parrish JE, Oostra BA, Verkerk AJ et al. Isolation of a GCC repeat showing expansion in FRAXF, a fragile site distal to FRAXA and FRAXE. Nat Genet 1994; 8:229-235.

87. Jones C, Penny L, Mattina T et al. Association of a chromosome deletion syndrome with a fragile site within the proto-oncogene CBL2. Nature 1995; 376:145-149.

88. Nancarrow JK, Kremer E, Holman K et al. Implications of FRA16A structure for the mechanism of chromosomal fragile site genesis. Science 1994; 264:1938-1941.

89. Brais B, Bouchard JP, Xie YG et al. Short GCG expansions in the PABP2 gene cause oculopharyngeal muscular dystrophy. Nat Genet 1998; 18:164-167.

90. Brock GJ, Anderson NH, Monckton DG. Cis-acting modifiers of expanded CAG/CTG triplet repeat expandability: associations with flanking GC content and proximity to CpG islands. Hum Mol Genet 1999; 8:1061-1067.

91. Manley K, Shirley TL, Flaherty L et al. MSH2 deficiency prevents in vivo somatic instability of the CAG repeat in Huntington disease transgenic mice. Nat Genet 1999; 23:471-473.

92. Lafreniere RG, Rochefort DL, Chretien N et al. Unstable insertion in the 5' flanking region of the cystatin B gene is the most common mutation in progressive myoclonus epilepsy type1, EPM1. Nat Genet 1997; 15:298-302.

93. Lalioti MD, Scott HS, Buresi C et al. Dodecamer repeat expansion in cystatin B gene in progressive myoclonus epilepsy. Nature 1997; 386:847-851.

94. Virtaneva K, D'Amato E, Miao J et al. Unstable minisatellite expansion causing recessively inherited myoclonus epilepsy, EPM1. Nat Genet 1997; 15:393-396.

95. Bois P, Jeffreys AJ. Minisatellite instability and germline mutation. Mol Cell Life Sci 1999; 55:1636-1648.

96. Gharani N, Waterworth DM, Batty S et al. Association of the steroid synthesis gene CYP11a with polycystic ovary syndrome and hyperandrogenism. Hum Mol Genet 1997; 6:397-402.

97. Mooser V, Mancini FP, Bopp S et al. Sequence polymorphisms in the apo(a) gene associated with specific levels of Lp(a) in plasma. Hum Mol Genet 1995; 4:173-181.

98. Xu W, Liu L, Emson PC et al. Evolution of a homopurine-homopyrimidine pentanucleotide repeat sequence upstream of the human inducible nitric oxide synthase gene. Gene 1997; 204:165-170.

99. Wang YH, Griffith JD. The [(G/C)3NN]n motif: a common DNA repeat that excludes nucleosomes. Proc Natl Acad Sci USA 1996; 93:8863-8867.

100. Bidichandani SI, Ashizawa T, Patel PI. The GAA triplet-repeat expansion in Friedreich ataxia interferes with transcription and may be associated with an unusual DNA structure. Am J Hum Genet 1998; 62:111-121.

101. Ohshima K, Montermini L, Wells RD et al. Inhibitory effects of expanded GAA.TTC triplet repeats from intron I of the Friedreich ataxia gene on transcription and replication in vivo. J Biol Chem 1998; 273:14588-14595.

102. Sakamoto N, Chastain PD, Parniewski P et al. Sticky DNA: self-association properties of long GAA. TTC repeats in R.R.Y triplex structures from Friedreich's ataxia. Mol Cell 1999; 3:465-475.

103. Klesert TR, Otten AD, Bird TD et al. Trinucleotide repeat expansion at the myotonic dystrophy locus reduces expression of DMAHP. Nat Genet 1997; 16:402-406.

104. Klesert TR, Cho DH, Clark JI et al. Mice deficient in six5 develop cataracts: implications for myotonic dystrophy. Nat Genet 2000; 25:105-109.

105. Korade-Mirnics Z, Tarleton J, Servidei S et al. Myotonic dystrophy: tissue-specific effect of somatic CTG expansions on allele-specific DMAHP/SIX5 expression. Hum Mol Genet 1999; 8:1017-1023.

106. Sarkar PS, Appukuttan B, Han J et al. Heterozygous loss of six5 in mice is sufficient to cause ocular cataracts. Nat Genet 2000; 25:110-114.

107. Philips AV, Timchenko LT, Cooper TA. Disruption of splicing regulated by a CUG-binding protein in myotonic dystrophy. Science 1998; 280:737-41.

108. Liquori CL, Ricker K, Moseley ML, Jacobsen JF, Kress W, Naylor SL, Day JW, Ranum LP. Myotonic dystrophy type 2 caused by a CCTG expansion in intron 1 of ZNF9. Science. 2001;293:864-867.

109. Ashizawa T, Dubel JR, Dunne PW et al. Anticipation in myotonic dystrophy. II. Complex relationships between clinical findings and structure of the GCT repeat. Neurology 1992; 42:1877-1883.

110. Rasmussen A, Matsuura T, Ruano L, Yescas P, Ochoa A, Ashizawa T, Alonso E. Clinical and genetic analysis of four Mexican families with spinocerebellar ataxia type 10. Ann Neurol. 2001;50:234-239.

111. Matsuura T, Ranum LP, Volpini V, Pandolfo M, Sasaki H, Tashiro K, Watase K, Zoghbi HY, Ashizawa T. Spinocerebellar ataxia type 10 is rare in populations other than Mexicans. Neurology. 2002;58:983.

112. Fujigasaki H, Tardieu S, Camuzat A, Stevanin G, LeGuern E, Matsuura T, Ashizawa T, Durr A, Brice A. Spinocerebellar ataxia type 10 in the French population. Ann Neurol. 2002;51:408-409.

THE MOLECULAR BASIS
OF FRIEDREICH ATAXIA

Massimo Pandolfo

Friedreich ataxia (FRDA) is the most common of the early-onset hereditary ataxias in Indo-European and North African populations. The disease was first described in 1863 by Nicholaus Friedreich, Professor of Medicine in Heidelberg. Friedreich's papers reported the essential clinical and pathological features of the disease, a "degenerative atrophy of the posterior columns of the spinal cord" leading to progressive ataxia, sensory loss and muscle weakness, often associated with scoliosis, foot deformity and heart disease. However, the subsequent description of atypical cases and of clinically similar diseases clouded classification for many years. Diagnostic criteria were established in the late 1970s, after a renewed interest in the disease prompted several rigorous clinical studies. The Québec Collaborative Group described the typical features of the disease in well-established cases.[1] Harding modified some of the Québec Collaborative Group diagnostic criteria to include cases at an early stage of the disease.[2] According to Harding, essential clinical features include:

i) autosomal recessive inheritance,
ii) onset before 25 years of age,
iii) progressive limb and gait ataxia,
iv) absent tendon reflexes in the legs,
v) electrophysiologic evidence of axonal sensory neuropathy, followed within five years of onset by: dysarthria, areflexia at all four limbs, distal loss of position and vibration sense, extensor plantar responses and pyramidal weakness of the legs.

The associated neuropathology is characterized by atrophy of the sensory pathways, with early loss of large neurons in the dorsal root ganglia (DRG), sensory axonal neuropathy, and degeneration of the posterior columns of the spinal cord. The cerebellum shows atrophy of the deep dentate nucleus, but its cortex is relatively preserved.[3]

Centre Hospitalier de lé Université de Montréal, Hopital Notre-Dame, 1560 rue Sherbrooke Est, Montréal, Québec H2L 4M1 Canada.

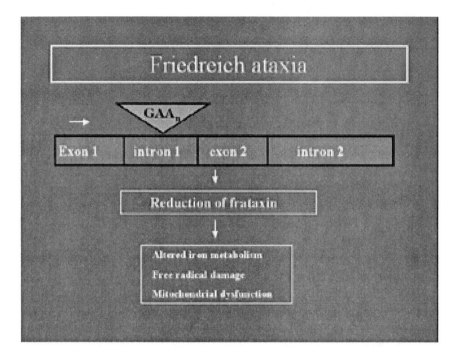

Figure 1. A model describing molecular pathways responsible for Friedreich ataxia (see text).

The identification of the FRDA gene and of its most common mutation, the unstable hyperexpansion of a GAA triplet repeat sequence (TRS),[4] has allowed to re-evaluate these issues on the basis of the results of molecular testing. While the above criteria certainly identify the typical cases of FRDA, it is now clear that the disease shows a remarkable clinical variability, sometimes even within the same sibship, a rather uncommon finding for recessive disorders. Variability involves age of onset, rate of progression, severity and extent of disease involvement.[5] Cardiomyopathy, kyphoscoliosis, pes cavus, optic atrophy, hearing loss and diabetes mellitus only occur in some patients. Atypical cases with an overall FRDA-like phenotype but missing some of the essential diagnostic features can be identified. These include late-onset Friedreich ataxia (LOFA), which develops after the age of 25, somatimes as late as the sixth decade, and Friedreich ataxia with retained tendon reflexes (FARR). The molecular basis for such a variability is still uncertain. Germline and somatic instability of the GAA TRS certainly plays a role, but additional genetic and environmental factors are clearly involved.[5]

GENE STRUCTURE AND EXPRESSION

The FRDA locus is in the proximal long arm of chromosome 9.[6] The gene contains seven exons spanning 95 Kb of genomic DNA. It is transcribed in the cen->tel direction. An unmethylated CpG island is in the first exon, a common finding at the 5' end of many genes. The major, and probably only functionally relevant mRNA, has a size of 1.3 Kb, corresponding to the first five exons, numbered 1 to 5a. The encoded protein, predicted to contain 210 aminoacids, was called frataxin.[4]

The gene is expressed in all cells, but at variable levels in different tissues and during development.[4,7,8] In adult humans, frataxin mRNA is most abundant in the heart and spinal cord, followed by liver, skeletal muscle, and pancreas. In mouse embryos, expression starts in the neuroepithelium at embryonic day 10.5 (E10.5), then reaches its highest level at E14.5 and into the postnatal period.[7,8] In developing mice, the highest levels of frataxin mRNA are found in the spinal cord, particularly at the thoracolumbar level, and in the dorsal root ganglia. The developing brain is also very rich in frataxin mRNA, which is abundant in the proliferating neural cells in the periventricular zone, in the cortical plates, and in the ganglionic eminence (precursor of the basal ganglia). Robust expression is also detected in the heart, in the axial skeleton, and in some epithelial (skin, teeth)[7] and mesenchymal tissues (brown fat).[8] In the adult mouse brain the level of frataxin mRNA is reduced and mostly confined to the ependyma, but remains high in the spinal cord and dorsal root ganglia.[8] Interestingly, protein levels (estimated by western blot analysis) remain high in the adult human and mouse brain and cerebellum.

Overall, frataxin expression is generally higher in mitochondria-rich cells, as cardiomyocytes and neurons. There is, however, a still unexplained additional cell specificity, which in the nervous system is reflected in a higher abundance of frataxin in specific neuronal types, as primary sensory neurons, and in specific developmental stages.

The GAA triplet repeat Mutation

The most common mutation causing Friedreich ataxia (98%) is the hyperexpansion of a GAA triplet repeat in the first intron of the frataxin gene (Fig. 1).[4] Disease-associated repeats contain from ~70 to more than 1,000 triplets, most commonly 600-900. Because of the recessive nature of the disease, affected individuals have expansions in both homologues of chromosome 9, while heterozygous carriers are clinically normal. This is the most common disease-causing triplet repeat expansion identified so far, 1 in 90 Europeans being a carrier.[9] Repeats in normal chromosomes contain up to ~40 triplets, 90 to >1,000 triplets in Friedreich's ataxia chromosomes.[4,10] No other disease has been recognized to date to be caused by an expansion of GAA•TTC. Occasional patients (~5%) are heterozygous for a GAA•TTC expansion and a missense or nonsense point mutation disrupting the frataxin coding sequence.[4,11] No patients have been identified so far that carry point mutations in both copies of the frataxin gene.

Instability of Expanded Repeats

The Friedreich's ataxia-associated expansion shows instability when transmitted from parent to child.[4,5,12,13] Expansions and contractions of expanded GAA repeats can both be observed. E alleles are equally likely to further expand or contract during maternal transmission, but most often contract during paternal transmission,[14,15] a result also supported by sperm analysis.[14] In this regard, Friedreich's ataxia resembles the other diseases associated with very large expansions in noncoding regions, as fragile X and myotonic dystrophy, while smaller expansions of CAG repeats in coding regions, found in dominant ataxias or Huntington disease, are more likely to undergo size increases during paternal transmission.

Mitotic instability, leading to somatic mosaicism for expansion sizes, can be observed in Friedreich's ataxia.[5] Analysis of GAA expansions reveals ample variations in different cell types or tissues from the same patient. Furthermore, heterogeneity among cells occurs at a variable degree in different tissues. For instance, cultured fibroblasts and cerebellar cortex show very little heterogeneity in expansion sizes among cells, lymphocytes are more heterogeneous, and most brain regions show a quite complex pattern of allele sizes, indicating extensive cellular heterogeneity.[16] While some of these differences could be accounted for by a major period of instability during the first weeks of embryonic development, GAA expanded repeats may be inherently more stable in some cell types.[16] In general, it is clear that determining the size of a patient's expansions in peripheral blood lymphocytes, from which DNA is usually obtained, only provides a single sample of the overall repeat size distribution occurring within that patient, and therefore only an approximate estimate of expansion sizes in affected tissues.

Origin and Mechanisms of Expansion of the Repeat

The GAA repeat associated with Friedreich's ataxia is localized within an Alu sequence (GAA-Alu). Alu sequences are a heterogeneous group of primate-specific interspersed repetitive DNA elements with an estimated frequency of 500,000 to 1 million copies per genome. They may serve as functional polIII genes and are probably derived from 7SL genes. Their pervasiveness and variability are the result of constant amplification and retrotranposon-mediated reinsertion throughout the genome over 65 million years of primate evolution.[17] Despite their diversity, Alu sequences can be grouped into subfamilies whose members share a few, common diagnostic base changes. By comparing differences between these sequences, Alu elements can be used as molecular clocks to estimate the age of a particular subfamily or member of a subfamily. GAA-Alu is assigned to the AluSx subfamily. Identity between GAA-Alu and the AluSx consensus sequence is 89%, in agreement with the overall 92% ± 3 identity between individual AluSx subfamily sequences and the consensus sequence. According to similarity calculation, the average age of the AluSx subfamily has been estimated at 37 million years[17]. The Friedreich's ataxia-associated GAA repeat lies in the middle of GAA-Alu, preceded by a stretch of an average of 16 A's, apparently derived from an expansion of the canonical A_5TACA_6 sequence linking the two halves of Alu sequences. GAA-Alu is flanked by a 13 bp perfect direct repeat (AAAATGGATTTCC), suggesting a recent Alu retroposition/insertion event, an idea supported by the estimated age of the AluSx subfamily.[18]

Alleles at the GAA repeat site can be subdivided into 3 classes depending on their length: short normal alleles, with 5-10 GAA triplets (SN, ~82% in Europeans); long normal alleles, with 12-60 GAA triplets (LN, ~17% in Europeans); disease associated expanded alleles, with >66 and up to 1,600 GAA triplets (E, ~1% in Europeans).[9,10]

The length polymorphism of the GAA repeat in normal alleles suggests that it was generated by two types of events. Small changes, plus or minus one trinucleotide, may have caused limited size heterogeneity. Such small changes were likely to be the consequence of occasional events of polymerase "stuttering" during DNA replication, i.e., slippage followed by mis-realignment of the newly synthesized strand by one or, rarely, a few repeat units.[19] This basic polymorphism-generating mechanism has been postulated for all simple-sequence repeats.[20] By comparison, the jump from the SN to the LN group was probably a singular event. Linkage disequilibrium (LD) studies carried out in European, but also Yemenite and North-African families, with single nucleotide polymorphisms spanning the frataxin gene (FAD1, ITR4, ITR3, and CS2) indicate a common origin of all chromosomes with alleles containing more than 12 GAA triplets. Essentially all these alleles share the same major haplotype or a minor, related haplotype that can be derived by one or two recombinations.[9] Possibly, the event that created LN alleles was the sudden duplication of an SN allele containing 8 or 9 GAA triplets, creating an LN allele with 16 or 18 GAA triplets. This occurred presumably in Africa, leading to a population of chromosomes with LN alleles sharing the same background haplotype. Single repeat insertion/deletions, resulting from DNA polymerase "stuttering", gave rise to the spectrum of stable GAA repeats ranging from 12 to about 25 triplets. One or a few of these chromosomes subsequently migrated to Europe and/or to the Middle East, but not to East Asia, where no LN (or E) alleles are found. It is hard to speculate about the mechanism leading to such a sudden doubling of the repeat, however similar events have been shown to occur in triplet repeats cloned into bacterial plasmids.[21] Recombination-based mechanisms as unequal sister-chromatid exchange and gene conversion have been proposed as generators of variability in VNTRs[20] and in microsatellites,[22] but alternative hypotheses such as the occurrence of an exceptionally large slippage event cannot be excluded.

The passage from LN to E alleles probably involved a second genetic event of the same kind, that generated "very long" LN alleles containing 32-36 GAA triplets still on the same haplotype background as the "shorter" LN alleles from which they derived. By reaching the instability threshold, estimated as 34 GAA triplets,[10] they form a reservoir for expansions. The occurrence of a second duplication event is suggested by the lack of both E and LN alleles with more than 21 GAA triplets alleles in Africans. The ethnic-geographic distribution of Friedreich's ataxia could be explained if the second event occurred prior to the divergence of Indo-Europeans and Afro-Asiatic speakers. According to the above scenario, the extent of LD between LN alleles and linked marker loci on chromosomes of African descent is expected to be lower than between LN and E alleles and the same marker in Europeans,[23,24] as in fact observed.[25] Accordingly, LN chromosomes in Africa appear to be 3.2 times older than the LN chromosomes in Europe, and these appear to be 1.27 times older than E chromosomes. Assuming the age of LN African chromosomes in the range of 100,000 years, one would date the origin of European LN chromosomes at about 30,000 years ago and that of the E chromosomes at about 25,000 years ago, i.e., following the Upper Paleolithic population expansion.[26]

It was possible to directly observe the hyperexpansion of premutant "very long" LN alleles containing more than 34 GAA triplets. This length is close to the instability threshold for other triplet repeat associated disorders, such as those involving CGG and CAG repeats.[27] Strand displacement during DNA replication is thought to be the mechanism that leads to reiterative synthesis and expansion.[28] For this phenomenon to occur, the displaced strand has to form some kind of secondary structure.[28] Although some authors have dismissed the possibility for a GAA strand to form a secondary hairpin structure,[28] this may be possible by A•A and G•G mismatches, which have been shown to occur under several conditions.[29,30] Moreover, a single DNA strand containing a GAA repeat is also able to form different types of secondary structure,[31] which may be involved in instability. A single CTT strand seems structureless,[31] and this difference may play a role in determining whether deletions or expansions are favored according to the direction of the replicating fork. Finally, strand displacement is promoted by stalling of DNA polymerase caused by an alternate DNA structure, or by tightly bound proteins, or both.[20] The triplex-forming ability of long Friedreich's ataxia GAA repeats, discussed below, may be involved in repeat instability by causing DNA polymerase stalling as well as by forming a target for protein binding.

Sequence Variants

A few LN alleles, and even some alleles in the full expansion range are interrupted by A to G transitions that create GAG or GGA triplets. These interruptions seem to prevent instability[32,33] and also render longer alleles non-pathogenic,[34] possibly by interfering with the ability of the repeat to adopt a secondary structure, as detailed below. The functional effect, if any, of commonly encountered stretches of 3 to 5 A nucleotides interrupting the regular run of GAA triplets is less clear.

Pathogenic Mechanisms: Triplexes and Sticky DNA

The current explanation for the observed inhibition of gene expression caused by long GAA repeats is that these sequences adopt a specific secondary structure that impedes transcription. This structure is most likely a triplex. Triplexes are three-stranded nucleic acid structures (usually DNA) formed at tracts of oligopurines (R) and oligopyrimidines (Y).[20,30,35-39] The third strand occupies the major grove of the

DNA double helix forming Hoogsteen pairs between R or Y bases with purines of the Watson-Crick base pairs. Intermolecular triplexes are formed between oligo- or polynucleotides (DNA or RNA) and target R•Y sequences on duplex DNA. Intramolecular triplexes are folded structures in supercoiled DNAs.[30,36-38] Triplexes were shown to exist in vivo.[30,36-39] According to the Intramolecular triplexes are formed at mirror repeat sequences at pH values below 7 when the third strand contains a C residue, due to the requirement for protonation.[30,36-38] Since thorough investigations were conducted in the 1980s on triplexes, substantial information is available on the effect of sequence and the type of R•Y sequences required, the effects of pH and methylation of C residues, the types of bi-triplexes (nodule DNA and sticky DNA) formed, the effect of interposing non-R•Y sequences, the influence of environmental factors on the stabilization of the four triplex isomers, the effect of stabilization by intercalating agents, and related factors.[30,36-50] Y•R•R triplexes are more versatile than Y•R•Y triplexes since they will tolerate more diverse pairing schemes and since their stability does not depend on lower pH but depends on the presence of divalent metal ions. We hypothesized[50] that the mechanism of reduction of abundance of mature frataxin mRNA in individuals with Friedreich's ataxia is the formation of an intermolecular triplex between the GAA•TTC in the first frataxin intron and the RNA segment with the GAA tract removed by splicing. Prior work[36,51,52] showed that the presence of a triplex inhibits transcription. In the case of long rGAA tracts (100 or more repeats) from Friedreich's ataxia cases, the triplex may be sufficiently stable thermodynamically to cause the reduction in abundance of the Friedreich's ataxia mature mRNA, whereas for shorter rGAA stretches from normal individuals (6-20 repeats), the triplex may be unstable and will not cause an inhibition. This hypothesis is consistent with the clinical observations that patients with longer GAA•TTC repeats (350-600 repeats) are more severely afflicted than patients with shorter repeats (150-250 repeats).[53]

We analyzed the effect of intronic GAA•TTC repeats on gene expression by transfecting COS-7 cells with constructs harboring GAA•TTC repeats of different lengths and orientations in an intron of a reporter gene. When $(GAA)_n$ was in the transcripts, as is the case in the frataxin gene, transcription and expression of the reporter gene were reduced proportionally to the repeat length. Repeats containing more than 33 triplets, close to the upper limit for normal alleles of the frataxin TRS,[2,54,55] started to inhibit gene expression. No increase in unspliced or partially spliced transcript was observed, suggesting that a defect in RNA splicing caused by the expanded GAA•TTC repeat, proposed as a cause of reduced frataxin gene expression in FRDA,[56] is unlikely. Along with the observation that transcription initiation is probably not affected, as suggested by RNase protection experiments, the occurrence of a transcriptional block at the repeat seems to be the most likely explanation for reduced gene expression. According to our observations, such a block is orientation-dependent, occurring only with transcription of GAA-containing RNA. Such purine-specific inhibition is in agreement with previous in vitro studies of pur•pyr sequences,[57-60] which indicated that under physiological conditions pur•pur•pyr triplex structures are preferentially formed and in vitro transcription of purine-rich RNA is specifically reduced.

Hence, these in vivo studies revealed that expanded GAA•TTC repeats from Friedreich's ataxia intron 1 inhibit transcription rather than posttranscriptional RNA processing. These data are consistent with prior results[61] on recombinant plasmids containing different lengths (9, 45, 79 and 100 repeats in length) using both pro-caryotic and eukaryotic RNA polymerases. This inhibition of transcription was most pronounced in the physiological orientation of transcription, when synthesis of the GAA-rich transcript was attempted. These investigators[61] hypothesized that the GAA•TTC repeat sequence adopts an unusual structure adding strong credibility to the concept of the involvement of triplexes in the pathology of Friedreich's ataxia.

Interestingly, essentially all workers in the Friedreich's ataxia field involved with these molecular biological processes have hypothesized or provided evidence for the involvement of triplexes in the disease etiology.[50,61-65] However, not all work-ers agree on the type of triplex formed. Griffin et al[66-68] suggested that the underly-ing molecular mechanism is the formation of an intermolecular RNA•DNA hybrid triplex structure. Grabczyk and Fishman[69] proposed instead that purine-rich RNA may bind to the single pyrimidine-rich DNA strand generated by the formation of an intramolecular DNA triplex, resulting in its stabilization. According to this model, a wave of negative supercoiling following transcription would trigger intramolecu-lar DNA triplex formation. In any case, the GAA-rich transcript would participate in stabilizing the structure, interfering with RNA elongation and preventing further transcription.

A new type of DNA structure, that implies intramolecular triplex formation, was shown to be adopted by lengths of GAA•TTC as found in Friedreich's ataxia. This structure was called "sticky DNA" and is formed by the association of two R•R•Y triplexes in plasmids containing long tracts of GAA•TTC. Sticky DNA was discovered as an anomalously retarded band in agarose gels in which linearized plasmids containing GAA•TTC were separated. Such slow-migrating band was shown to have a number of physicochemical properties that are typical of intramolcular R•R•Y triplexes. In particular, the retarded band appeared only if the plasmid was negatively supercoiled prior to linearization, and it was sensitive to divalent ion concentration and temperature as is typical for R•R•Y triplexes. The possible intermolecular nature of the structure was suggested by the correlation be-tween its abundance and plasmid DNA concentration. This was proven by electron microscopy analysis, that revealed bimolecular complexes formed by joining two plasmids through the region containing the GAA•TTC TRS. An excellent correla-tion was found between the lengths of GAA•TTC and the formation of this novel conformation: Friedreich's ataxia patients have 66 or more repeats,[53] sticky DNA was found only for repeats longer than 59 units. As these data suggest a role of this structure in the pathogenesis of Friedreich's ataxia, we recently carried out in vitro transcription studies of (GAA•TTC)$_n$ repeats (where n=9 to 150) using T7 or SP6 RNA polymerase. When a gel-isolated sticky DNA template was transcribed, the amount of full-length RNA synthesized was significantly reduced compared to the transcription of the linear template. Surprisingly, transcriptional inhibition was observed not only for the sticky DNA template but also another DNA molecule used as an internal control in an orientation independent manner. The molecular mechanism of transcriptional inhibition by sticky DNA was a sequestration of the RNA polymerases by direct binding to the complex DNA structure. These results further support role of sticky DNA in Friedreich's ataxia and suggest that it may include the sequestration of transcription factors.

We observed that a $(GAAGGA \cdot TCCTTC)_{65}$ sequence, also found in intron 1 of the frataxin gene, does not form sticky DNA nor inhibit transcription in vivo and in vitro nor associate with the Friedreich's ataxia disease state.[34] This finding suggests that interruptions in the $GAA \cdot TTC$ sequence may destabilize its structure and facilitate transcription. Two recent findings by our laboratories support this hypothesis, that is central to our proposal. First, a systematic analys analysis of the effects of introducing interruptions into a $(GAA \cdot TTC)_{150}$ repeat by substituting an increasing number of As with Gs has confirmed that the sticky DNA/triplex structure is progressively destabilized and it fails to form when the sequence becomes $(GAAGGA \cdot TCCTTC)_{75}$. As the tendency to form a sticky DNA/triplex structure decreases, less and less inhibition of transcription is observed in vivo and in vitro.

Genotype-Phenotype Correlation for the GAA Expansion

As expected by the experimental finding that smaller expansions allow a higher residual gene expression,[63,70,71] expansion sizes have an influence on the severity of the phenotype. A direct correlation has been firmly established between the size of GAA repeats and earlier age of onset, earlier age when confined in wheelchair, more rapid rate of disease progression, and presence of non-obligatory disease manifestations indicative of more widespread degeneration.[5,12,13,15,73,74] However, differences in GAA expansions account for only about 50% of the variability in age of onset, indicating that other factors influence the phenotype. These may include somatic mosaicism for expansion sizes, variations in the frataxin gene itself, modifier genes and environmental factors.

POINT MUTATIONS

About 2% of the Friedreich ataxia chromosomes carry GAA repeat of normal length, but have a missense, nonsense, or splice site mutations ultimately affecting the frataxin coding sequence.[4,11,75] All affected individuals with a point mutation so far identified are heterozygous for an expanded GAA repeat on the other homologue of chromosome 9. It is possible that homozygotes for point mutations have not yet been found just because point mutations are rare, but it is more likely that homozygosity for frataxin point mutations would cause a lethal phenotype, as suggested by the recent observation that frataxin knock-out mice[76] and mice homozygous for a frataxin missense mutation (P. Ioannu, personal communication) die during embryonic development.

A few missense mutations are associated with milder atypical phenotypes with slow progression, suggesting that the mutated proteins preserve some residual function. Patients carrying the G130V mutation have early onset but slow progression, no dysarthria, mild limb ataxia, and retained reflexes.[11,75] A similar phenotype occurs in individuals with the mutations D122Y[11] and R165P.[77] For unclear reasons, optic atrophy is more frequent in patients with point mutations of any kind (50%).[11]

FRATAXIN STRUCTURE AND FUNCTION

Subcellular Localization

Frataxin does not resemble any protein of known function. It aminoacid sequence does not predict any transmembrane domain. It is highly conserved during evolution,[4] with homologs in mammals, invertebrates, yeast and plants. The protein is targeted to the mitochondria,[70,78,79] as first discovered by observing the intracellular localization of frataxin-green fluorescent protein (GFP) fusion proteins.[78,80] The mitochondrial localization of endogenous frataxin was then demonstrated by immunocytofluorescence, western blot analysis of cellular fractions obtained by differential centrifugation, and immunoelectron microscopy (EM).[70] The protein was subsequently localized to the mitochondrial matrix.[79]

Frataxin has an N-terminal mitochondrial targeting sequence, which is proteolytically removed by the mitochondrial processing peptidase (MPP) after the protein is imported into mitochondria. According to some authors, MPP first removes 40 aminoacids, then about 20 more aminoacids in a second proteolytic step,[79] according to others cleavage occurs in only one step.[81] At least in the case of the yeast homologous protein, frataxin maturation was shown to be promoted by a specific mitochondrial heat-shock protein of the hsp70 class, *ssq1p* . Yeast mutants with a defect of *ssq1p* process frataxin slowly and accumulate iron in mitochondria as frataxin knock-out mutants do (see below).[82]

The Yeast Model

Genes can be easily disrupted (knocked out) in yeast by homologous recombination, providing a powerful tool to study their function. This was accomplished for the yeast frataxin homolog gene (YFH1). Most YFH1 knock-out yeast strains, called *ΔYFH1*, lose the ability to carry out oxidative phosphorylation, forming *petite* colonies with defects or loss of mitochondrial DNA that cannot grow on non-fermentable substrates.[78,83] In *ΔYFH1*, iron accumulates in mitochondria, more then 10-fold in excess of wild type yeast, at the expense of cytosolic iron. Loss of respiratory competence requires the presence of iron in the culture medium, and occurs more rapidly as iron concentration in the medium is increased, suggesting that permanent mitochondrial damage is the consequence of iron toxicity.[8] Iron in mitochondria can react with reactive oxygen species (ROS) that form in these organelles. Even in normal mitochondria, a few electrons prematurely leak from the respiratory chain, mostly from reduced ubiquinone (or probably its semiquinone form), directly reducing molecular oxygen to superoxide (O_2^-). Mitochondrial Mn-dependent superoxide dismutase (SOD2) generates hydrogen peroxide (H_2O_2) from O_2^-, then glutathione peroxidase oxidizes glutathione to transforms H_2O_2 into H_2O. Iron may intervene in this process and be engaged in a cycle with O_2^- and H_2O_2 as follows:

$$Fe(III) + O_2^- + 2H^+ \rightarrow Fe(II) + H_2O_2$$
$$Fe(II) + H_2O_2 \rightarrow Fe(III) + OH^\bullet + OH^- \text{ (Fenton reaction)}$$

The hydroxyl radical ($OH^\bullet$) produced by the Fenton reaction is highly toxic and causes lipid peroxidation, protein and nucleic acid damage. Occurrence of the Fenton reaction in $\Delta YFH1$ yeast cells is suggested by their highly enhanced sensitivity to H_2O_2.[78]

Disruption of frataxin causes a general dysregulation of iron metabolism in yeast cells. Because iron is trapped in the mitochondrial fraction, a relative deficit in cytosolic iron results, causing a marked induction (10- to 50-fold) of the high-affinity iron transport system of the cell membrane, normally not expressed in yeast cells that are iron replete.[78] As a consequence, iron crosses the plasma membrane in large amounts and further accumulates in mitochondria, engaging the cell in a vicious cycle.

The reason why $\Delta YFH1$ cells accumulate iron in the mitochondrial fraction may in principle involve increased iron uptake, altered utilization or decreased export from these organelles. Experiments involving induction of frataxin expression from a plasmid transformed into $\Delta YFH1$ yeast cells indicate that the protein stimulates a flux of non-heme iron out of mitochondria,[84] but the mechanism and the involved transporter remain obscure. Interestingly, heme synthesis is normal in $\Delta YFH1$ yeast, suggesting that ferrochelatase function and the transport of heme out of mitochondria are not affected by frataxin deficiency.

The possibility of interpreting the current experimental data in different ways, leaves open the question of the primary function of frataxin, even in yeast. Consequently, it is not yet possible to state whether mitochondrial damage is entirely the consequence of free radicals, or it is in part a direct consequence of the missing primary activity of frataxin. Several mitochondrial enzymes are known to be impaired in $\Delta YFH1$ yeast cells, particularly respiratory chain complexes I, II, and III and aconitase.[85] These enzymes all contain iron-sulfur (Fe-S) clusters in their active sites. Fe-S clusters are remarkably sensitive to free radicals,[86] so a deficit can be reasonably ascribed to oxidative damage. However, a specific synthetic pathway has been recently discovered for Fe-S clusters in yeast mitochondria.[87] Remarkably, defects in several enzymes in the pathway lead to mitochondrial iron accumulation, similar to what is observed in $\Delta YFH1$. This has prompted some researchers to suggest that yeast frataxin may itself be involved in Fe-S cluster synthesis. To date, the only direct piece of data that may be interpreted to support this hypothesis is that aconitase activity is still reduced in $\Delta YFH1$ cells to 50% of control cells when iron in the medium is very low and loss of mitochondrial function does not occur.[88]

A recent study even suggested that frataxin has no direct role in iron metabolism. By observing increased oxidative phosphorylation activity in adypocites that overexpress frataxin, those authors concluded that frataxin's main role is to stimulate mitochondrial function in a still unknown manner. Iron accumulation would non-specifically result from decreased mitochondrial activity.

Biochemical Studies

The yeast frataxin homolog, YFH1p (the protein product of the YFH1 gene), may be an iron-binding protein.[89] Monomers of YFH1p are not capable of binding iron, but experiments using gel filtration and analytical ultracentrifugation have suggested that a high molecular weight YFH1p-iron complex may form when ferrous iron is added to the protein at a 40:1 molar ratio. Small amount of intermediates containing 2,3 or more molecules of YFH1p complexed with iron form at lower iron:protein ratios. The high molecular weight complex would resemble ferritin, containing a large number of iron atoms within a proteinaceous shell made by frataxin.[89] Preliminary western blot analysis of gel filtration fractions of yeast extracts suggests that high molecular weight complexes containing YFH1p may exist in vivo. According to these data, YFH1p may protect iron in mitochondria from contacts with free radicals. Since iron in the complexes seems to be readily accessible to chelators, so probably bioavailable, YFH1p could be a sort of mitochondrial iron chaperone, in the absence of which several biosyntheses and transport processes are impaired and iron accumulates in a toxic, redox-active form. Unfortunately, a different group has not been able to replicate the frataxin-iron binding experiments and has reported that no binding can be detected at any frataxin:iron ratio.[90] While this question is not settled at this time, it remains critical for the understanding of frataxin function.

Protein Structure

The structure of frataxin is the object of intensive analysis. A recent publication[91] described the crystal structure of frataxin, a second one the NMR-derived structure of the soluble protein.[90] A third paper reported the crystal structure of the frataxin bacterial homolog, CyaY.[92] All studies agree that mature frataxin is a compact, globular protein containing an N-terminal α helix, a middle β sheet region composed of seven β strands, a second α helix, and a C-terminal coil. The α helices are folded upon the β sheet, with the C-terminal coil filling a groove between the two α helices. Hydrophobic aminoacids are clustered on the sides of the α helices and of the β sheet that form the central core of the structure. Many of these aminoacids are necessary for the stability of the structure and cannot be replaced, as demonstrated by their conservation in different species, and by the disruptive effect of mutations on frataxin stability. On the outside, some portions of the surface of frataxin are remarkably conserved. These include a ridge of negatively charged residues and a patch of hydrophobic residues. The size and nature of the conserved surface regions suggest that they interact with a large ligand, probably a protein. However, experiments aimed to identify a protein partner of frataxin, mostly by using the yeast two-hybrid method, have so far failed. Frataxin does not have any feature resembling known iron-binding sites. However, the negatively charged ridge confers some resemblance to a unique bacterial ferritin in which an iron-binding pouch is formed by two adjoining subunits. The crystal structure study demonstrated that iron only non-specifically binds the frataxin monomer. The NMR study failed to identify any structural change of soluble frataxin after iron addition. Hopefully, a more extensive correlation between structural data and biochemical findings will soon be available, and possibly help to solve the problem of iron binding.

CURRENT HYPOTHESES FOR THE PATHOGENESIS OF FRIEDREICH ATAXIA

Mitochondrial Iron Metabolism Dysfunction and Oxidative Damage

Normal human frataxin is able to complement the defect in $\Delta YFH1$ cells, while human frataxin carrying a point mutation found in Friedreich ataxia patients is unable to do so,[83] strongly suggesting that the function of YFH1p is conserved in human frataxin.

Several observations reinforce the hypothesis that altered iron metabolism, free radical damage, and mitochondrial dysfunction all occur in Friedreich ataxia (Fig. 1). Involvement of iron was suggested twenty years ago by the finding of deposits of this metal in myocardial cells from Friedreich ataxia patients.[93] Iron accumulation has been demonstrated by magnetic resonance imaging (MRI) in the dentate nucleus, a severely affected structure in the central nervous system.[94] We have confirmed an increase in dentate nucleus iron by atomic absorption spectroscopy analysis of pathological samples from three Friedreich ataxia patients (our unpublished data). The observation of a moderate, but significant increase in iron concentration in the mitochondrial fraction from Friedreich ataxia fibroblasts has been reported.[95] Oxidative stress is suggested by the observation that patients with Friedreich ataxia have increased plasma levels of malondialdheyde, a lipid peroxidation product,[96] and a product of oxidative damage to DNA. In addition, Friedreich ataxia fibroblasts are sensitive to low doses of H_2O_2, that induce cell shrinkage, nuclear condensation and apoptotic cell death at lower doses than in control fibroblasts[97] (and our unpublished observations). This finding suggests that even non-affected cells are in an "at risk" status for oxidative stress as a consequence of the primary genetic defect. A further hint of a possible role of free radicals comes from the observation that vitamin E deficiency produces a phenotype resembling Friedreich ataxia.[98] Vitamin E localizes in mitochondrial membranes where it acts as a free radical scavenger.[99]

Mitochondrial dysfunction has been proven to occur in vivo in Friedreich ataxia patients. Magnetic resonance spectroscopy analysis of skeletal muscle shows a reduced rate of ATP synthesis after exercise, which is inversely correlated to GAA expansion sizes.[100] Rötig et al[85] also demonstrated the same multiple enzyme dysfunctions found in *ΔYFH1* yeast (deficit of respiratory complexes I, II and III, and of aconitase) in endomyocardial biopsies from two Friedreich ataxia patients.[85] A general abnormality of iron metabolism may also be occurring in Friedreich ataxia patients, as suggested by the high level of circulating transferrin receptor, the principal carrier of iron into human cells,[101] which may reflect a relative cytosolic iron deficit as observed in the yeast model. In higher eukaryotes, cytosolic iron is sensed by two iron responsive element binding proteins (IRP-1 and IRP-2), that regulate the expression of several genes at the post-transcriptional level. When activated by low iron, they bind to specific sequence elements (iron responsive elements, IREs) present in some mRNAs. IRP binding stabilizes mRNAs encoding proteins that enhance iron uptake, as the transferrin receptor (TfR), while blocking the translation of mRNAs encoding proteins that utilize or store iron, as ferritin.[102] IRP-1 is a cytosolic aconitase containing an Fe-S cluster. It is activated not only in response to low cytosolic iron, but also to oxidative radicals and to signaling molecules as nitric oxide (NO) and carbon monoxide (CO).[102] If the loss of aconitase activity observed by Rötig et al[85] involves the cytosolic enzyme, it might result in changes in the abundance of IRP-1-regulated proteins,[85] including the observed increase in transferrin receptor. It should be noticed that the expression of frataxin does not seem to be regulated by iron (our unpublished observation) and its mRNA does not contain an IRE.

Frataxin, Cell Survival and Development

It is important to consider frataxin function in relation to development. This is a so far completely unexplored area. The generation of a frataxin knock-out mouse[76] has revealed that homozygous knock-out mice die as early as embryonic day 7 (E7). While total absence of frataxin leads to cell death in the early embryo, a reduced level of the protein, as observed in Friedreich ataxia patients, may only affect some cells that are dependent on a normal level of frataxin to survive through some critical step in their development. Sensory neurons in the dorsal root ganglia may be amongst these cells. They are lost very early in Friedreich ataxia, the loss seems to be non-progressive, and may be developmental.[3,103] When an animal model for the disease will be available, it will be worth exploring whether frataxin deficiency renders these cells particularly vulnerable to programmed cell death, and through what mechanism. Such studies may provide insight in the so far unexplained specific cell vulnerability of some sensory neurons to Friedreich ataxia.

Animal Models

The early embryonic lethality of frataxin ko mice has complicated the effort to generate an animal model of the disease. A viable mouse model has been obtained through a conditional gene targeting approach. A heart and striated muscle frataxin-deficient line and a line with more generalized, including neural, frataxin-deficiency have been generated.[104] These mice reproduce important progressive pathophysiological and biochemical features of the human disease: cardiac hypertrophy without skeletal muscle involvement in the heart and striated muscle frataxin-deficient line, large sensory neuron dysfunction without alteration of the small sensory and motor neurons in the more generalized frataxin-deficient line, deficient activities of complexes I-III of the respiratory chain and of the aconitases in both lines.[104] Time-dependent intramitochondrial iron accumulation occurs in the heart of the heart and striated muscle frataxin-deficient line.[104] These animals provide an important resource for pathophysiological studies and for testing of new treatments. However, they still do not mimic the situation occurring in the human disease because conditional gene targeting leads to complete loss of frataxin in some cells at a specific time in development, while Friedreich ataxia is characterized by partial frataxin deficiency in all cells and throughout life. Therefore, there is still a need to develop new animal models of the disease.

APPROACHES FOR TREATMENT

Based on the hypothesis that iron-mediated oxidative damage plays a major role in the pathogenesis of Friedreich ataxia, removal of excess mitochondrial iron and/or anti-oxidant treatment may in principle be attempted. However, removal of excess mitochondrial iron is problematic with the currently available drugs. Desferioxamine (DFO) is effective in chelating iron in the extracellular fluid and cytosol, not directly in mitochondria. Furthermore, DFO toxicity may be higher when there is no overall iron overload. Thus, chelation therapy has a number of unknowns: it is probably better tested in pilot trials involving a small number of closely monitored patients. Iron depletion by phlebotomy, though less risky, presents the same uncertainties concerning possible efficacy. As far as antioxidants are concerned, these include a long list of molecules with specific mechanisms of action and pharmacokinetic properties. To have the potential to be effective in FRDA, an antioxidant must protect against the damage caused by the free radicals involved in this disease, in particular $OH^{\bullet}$, act in the mitochondrial compartment and be able to cross the blood-brain barrier. At this time, CoQ derivatives, like its short chain analog idebenone, appear to be interesting molecules and are object of pilot studies.[105] However, new knowledge on frataxin function and pathogenesis is needed to progress towards an effective treatment of the disease. Pharmacological agents may be identified that counteract specific effects of frataxin deficiency. In the long term, gene replacement, protein replacement, or reactivation of the expression of endogenous frataxin could be cures and are all worth exploring.

ACKNOWLEDGMENTS

Work in the author's laboratory was supported by grants from the Medical Research Council of Canada, the National Institute of Neurological Diseases and Stroke (NINDS), and the Muscular Dystrophy Association (MDA), USA.

REFERENCES

1. Geoffroy G, Barbeau A, Breton, G et al. Clinical description and roentgenologic evaluation of patients with Friedreich ataxia. Can J Neurol Sci 1976; 3:279-286.
2. Harding AE. Friedreich ataxia: A clinical and genetic study of 90 families with an analysis of early diagnosis criteria and intrafamilial clustering of clinical features. Brain 1981; 104:589-620.
3. Koeppen A. The neuropathology of inherited ataxias. In: Manto M, Pandolfo M, eds. The Cerebellum and its Disorders. Cambridge: Cambridge University Press, 2001: Part VII, Neuropathology: 25.
4. Campuzano V, Montermini L, Moltó MD, Pianese L, Cossée M, Cavalcanti F et al. Friedreich ataxia: Autosomal recessive disease caused by an intronic GAA triplet repeat expansion. Science 1996; 271:1423-1427.
5. Montermini L, Richter A, Morgan K, Justice CM, Julien D, Castelloti B et al. Phenotypic variability in Friedreich ataxia: role of the associated GAA triplet repeat expansion. Ann Neurol 1997; 41:675-682.
6. Chamberlain S, Shaw J, Rowland A et al. Mapping of mutation causing Friedreich' ataxia to human chromosome 9. Nature 1988; 334:248-250.
7. Jiralerspong S, Liu Y, Montermini L, Stifani S, Pandolfo M. Frataxin shows developmentally regulated tissue-specific expression in the mouse embryo. Neurobiol Dis 1997; 4:103-113.
8. Koutnikova H, Campuzano V, Foury F, Dollé P, Cazzalini O, Koenig M. Studies of human, mouse and yeast homologues indicate a mitochondrial function for frataxin. Nat Genet 1997; 16:345-351.
9. Cossée M, Schmitt M, Campuzano V et al. Evolution of the Friedreich ataxia trinucleotide repeat expansion: Founder effect and premutations. Proc Natl Acad Sci USA 1997; 94:7452-7457.
10. Montermini L, Andermann E, Labuda M et al. The Friedreich ataxia GAA triplet repeat: premutation and normal alleles. Hum Mol Genet 1997; 6:1261-1266.
11. Cossée M, Dürr A, Schmitt M et al. Frataxin point mutations and clinical presentation of compound heterozygous Friedreich ataxia patients. Ann Neurol 1999; 45:200-206.
12. Dürr A, Cossée M, Agid Y et al. Clinical and genetic abnormalities in patients with Friedreich ataxia. N Engl J Med 1996; 335:1169-1175.
13. Filla A, De Michele G, Cavalcanti F et al. The relationship between trinucleotide (GAA) repeat length and clinical features in Friedreich ataxia. Am J Hum Genet 1996; 59:554-560.
14. Pianese L, Cavalcanti F, De Michele G, Filla A, Campanella G, Calabrese et al. The effect of parental gender on the GAA dynamic mutation in the FRDA gene. Am J Hum Genet 1997; 60:463-466.
15. Monros E, Moltó MD, Martinez F et al. Phenotype correlation and intergenerational dynamics of the Friedreich ataxia GAA trinucleotide repeat. Am J Hum Genet 1997; 61:101-110.
16. Montermini L, Kish SJ, Jiralerspong S, Lamarche JB, Pandolfo M. Somatic mosaicism for the Friedreich's ataxia GAA triplet repeat expansions in the central nervous system. Neurology 1997; 49:606-610.
17. Kapitonov V, Jurka J. The age of Alu subfamilies. J Mol Evol 1996; 42:59-65.
18. Arcot SS, Fontius JJ, Deininger PL and Batzer MA. Identification and analysis of a young polymorphic Alu element. Biochem et Biophys 1995; 1263:99-102.

19. Richards RI, Sutherland GR. Simple repeat DNA is not replicated simply. Nature Genet 1994; 6:114-116.

20. Wells RD. Molecular basis of genetic instability of triplet repeats. J Biol Chem 1996; 271:2875-2878.

21. Pluciennik A, Iyer RR, Parniewski P, Wells RD. Tandem duplication. A novel type of triplet repeat instability. J Biol Chem 2000; 275:28386-28397.

22. Jakupciak JP, Wells RD. Gene conversion (recombination) mediates expansions of CTG•CAG repeats. J Biol Chem 2000; 275:40003-40013.

23. Labuda D, Labuda M, Zietkiewicz E. The genetic clock and the age of the founder effect in growing populations: a lesson from French Canadians and Ashkenazim. Am J Hum Genet 1997; 61:768-771.

24. Harpending HC, Batzer MA, Gurven M et al. Genetic traces of ancient demography. Proc Natl Acad Sci 1998; 95:1961-1967.

25. Labuda M, Labuda D, Miranda C, Poirier J, Soong B, Barucha NE et al. Unique origin and specific ethnic distribution of the Friedreich ataxia GAA expansion. Neurology 2000; 54:2322-2324.

26. Geschwind DH, Perlman S, Grody W et al. The Friedreich's Ataxia GAA repeat expansion in patients with recessive or sporadic ataxia. Neurology 1997; 49:1004-1009.

27. Eichler EE, Holden JJA, Popovich BW, Reiss AL, Snow K, Thibodeau SN et al. Length of uninterrupted CGG repeats determines instability in the FMR1 gene. Nature Genet 1994; 8:88-94.

28. Gacy AM, Goellner G, Juranic N, Macura S, McMurray CT. Trinucleotide repeats that expand in human disease form hairpin structures in vitro. Cell 1995; 81:533-540.

29. Skelly JV, Edwards KJ, Jenkins TC, Neidle S. Crystal structure of an oligonucleotide duplex containing G•G base pairs: Influence of mispairing on DNA backbone conformation. Proc Natl Acad Sci USA 1993; 90:804-808.

30. Wells RD, Collier DA, Hanvey JC, Shimizu M, Wohlrab F. The chemistry and biology of unusual DNA structures adopted by oligopurine-oligopyrimidine sequences. FASEB J 1988; 2:2939-2949.

31. LeProust EM, Pearso CE, Sinden RR, Gao X. Unexpected formation of parallel duplex in GAA and TTC trinucleotide repeats of Friedreich's ataxia. J Mol Biol 2000; 302:1063-1080.

32. Imbert G, Kretz C, Johnson K, Mandel J-L. Origin of the expansion mutation in myotonic dystrophy. Nature Genet 1993; 3:72-75.

33. Altschul SF, Gish W, Miller W, Myers EW, Lipman DJ. Basic local alignment search tool. J Mol Biol 1990; 215:403-410.

34. Ohshima K, Sakamoto N, Labuda M et al. A nonpathogenic GAAGGA repeat in the Friedreich gene : Implications for pathogenesis. Neurol 1999; 53:1854-1857

35. Wells RD, Sinden RR. Defined ordered sequence DNA, DNA structure, and DNA-directed mutation. In: Davies K, Warren S, eds. Genome Analysis. Vol. 7. Genome Rearrangement and Stability. Cold Spring Harbor Press, 1993:107-138.

36. Frank-Kamenetskii MD, Mirkin SM. Triplex DNA Structures. Annu Rev Biochem 1995; 64:65-95.

37. Soyfer VN, Potaman VN. Triple-Helical Nucleic Acids. New York: Springer-Verlag Publishers, 1996.

38. Sinden RR. DNA Structure and Function. San Diego: Academic Press, Inc., 1994.

39. Guieysse A-L, Praseuth D, Grigoriev M, Harel-Bellan A and Helene C. Detection of covalent triplex with human cells. Nucl Acids Res 1996; 24:4210-4216.

40. Bacolla A, Ulrich MJ, Larson JE, Ley TJ, Wells RD. An intramolecular triplex in the human gamma-globin 5'-flanking region is altered by point mutations associated with hereditary persistence of fetal hemoglobin. J Biol Chem 1995; 270:24556-24563.

41. Xu G, Goodridge AG. Characterization of a polypyrimidine/polypurine tract in the promoter of the gene for chicken malic enzyme. J Biol Chem 1996; 271:16008-16019.

42. Hanvey JC, Klysik J, Wells RD. Influence of DNA sequence on the formation of non-B right-handed helices in oligopurine/oligopyrimidine inserts in plasmids. J Biol Chem 1988; 263:7386-7396.

43. Hanvey JC, Shimizu M, Wells RD. Intramolecular DNA triplexes in supercoiled plasmids. Proc Natl Acad Sci 1988; 85:6292-6296.

44. Shimizu M, Hanvey JC, Wells RD. Intramolecular DNA triplexes in supercoiled plasmids: I. Effect of loop size on formation and stability. J Biol Chem 1989; 264:5944-5949.

45. Hanvey JC, Shimizu M, Wells RD. Intramolecular DNA triplexes in supercoiled plasmids: II. Effect of base composition and non-central interruptions on formation and stability. J Biol Chem 1989; 264:5950-5956.

46. Hanvey JC, Shimizu M. Wells RD. Site-specific inhibition of EcoRI restriction/modification enzymes via DNA triple helix. Nucl Acids Res 1989; 18:157-161.

47. Shimizu M, Hanvey JC, Wells RD. Multiple Non-B-DNA Conformations of polypurine/polypyrimidine sequences in plasmids. Biochemistry 1990; 29:4704-4713.

48. Kang S, Wohlrab F, Wells RD. Metal ions cause the isomerization of certain intramolecular triplexes. J Biol Chem 1992; 267:1259-1264.

49. Kang S, Wohlrab F, Wells RD. GC rich flanking tracts decrease the kinetics of intramolecular DNA triplex formation. J Biol Chem 1992; 267:19435-19442.

50. Ohshima K, Kang S, Larson JE, Wells RD. Cloning, characterization, and properties of seven triplet repeat DNA sequences. J Biol Chem 1996; 271:16773-16783.

51. Morgan AR, Wells RD. Specificity of the three-stranded complex formation between double-stranded DNA and single-stranded RNA containing repeating nucleotide sequences. J Mol Biol 1968; 37:63-80.

52. Postel EH, Flint SJ, Kessler DJ, Hogan ME. Evidence that a triplex-forming oligodeoxyribonucleotide binds to the c-myc promoter in HeLa cells, thereby reducing c-myc mRNA levels. Proc Natl Acad Sci USA 1991; 88:8227-8231.

53. Pandolfo M, Koenig M. Friedreich's ataxia. In: Wells RD, Warren ST, eds. Genetic Instabilities and Hereditary Neurological Diseases. San-Diego: Academic Press Inc., 1998:373-400.

54. Said G, Marion MH, Selva J, Jamet C. Hypotrophic and dying-back nerve fibers in Friedreich's ataxia. Neurology 1986; 36:1292-1299.

55. Junck L, Gilman S, Gebarski SS, Koeppe RA, Kluin KJ, Markel DS. Structural and functional brain imaging in Friedreich's ataxia. Arch Neurol 1994; 51:349-355.

56. Palau F, De Michele G, Vilchez JJ, Pandolfo M, Monros E, Cocozza S et al. Early-onset ataxia with cardiomyopathy and retained tendon reflexes maps to Friedreich's ataxia locus on chromosome 9q. Ann Neurol 1997; 37:359-362.

57. Finocchiaro G, Baio G, Micossi P, Pozza G, Di Donato S. Glucose metabolism alterations in Friedreich's ataxia. Neurology 1998; 38:1292-1296.

58. Schoenle EJ, Boltshauser EJ, Baekkeskov S, Landin Olsson M, Torresani T, von Felten A. Preclinical and manifest diabetes mellitus in young patients with Friedreich's ataxia: No evidence of immune process behind the islet cell destruction. Diabetologia 1989; 32:378-381.

59. Fantus IG, Seni MH, Andermann E. Evidence for abnormal regulation of insulin receptors in Friedreich's ataxia. J Clin Endocrinol Metab 1993; 76:60-63.

60. Leone M, Rocca WA, Rosso MG, Mantel N, Schoenberg BS, Schiffer D. Friedreich's disease: Survival analysis in an Italian population. Neurology 1988; 38:1433-1438.

61. Bidichandani SI, Ashizawa T, Patel PI. The GAA triplet-repeat expansion in Friedreich's ataxia interferes with transcription and may be associated with an unusual DNA structure. Am J Hum Genet 1998; 62:111-121.

62. Filla A, De Michele G, Cavalcanti F, Pianese L, Monticelli A, Campanella G et al. The relationship between trinucleotide (GAA) repeat length and clinical features in Friedreich ataxia.. Am J Hum Genet 1996; 59:554-560.

63. Ohshima K, Montermini L, Wells RD, Pandolfo M. Inhibitory effects of expanded GAA•TTC triplet repeats from intron 1 of Friedreich's ataxia gene on transcription and replication in vivo. J Biol Chem 1998; 273:14588-14595.

64. Grabczyk E, Usdin K. Length dependent transcription attenuation in the Friedreich's ataxia triplet expansion mutation (GAA) via triple helix formation. Abstract presented at 17th International Congress of Biochemistry and Molecular Biology in conjunction with 1997 Annual Meeting of the American Society for Biochemistry and Molecular Biology, San Francisco, CA, August 24-29, 1997, Abstract No. 1998 (1997).

65. Gacy AM, Goellner GM, Spiro C, Dyer R, Mikesell M, Yao JZ et al. DNA structures associated with class I expansion of GAA in Friedreich's ataxia. Poster number CS3-103, presented at Santa Fe meeting on "Unstable Triplets, Microsatellites, and Human Disease," April 1-6, 1997 (J. Griffith, R.D. Wells, and D. L. Nelson, organizers).

66. Reaban ME, Griffin JA. Induction of RNA-stabilized DNA conformers by transcription of an immunoglobulin switch region. Nature 1990; 348:342-344.

67. Reaban ME, Griffin JA. Scientific correspondence. Nature 1991; 351:447-448.

68. Reaban ME, Lebowitz J, Griffin JA. Transcription induces the formation of a stable RNA.DNA hybrid in the immunoglobulin alpha switch region. J Biol Chem 1994; 269:21850-21857.

69. Grabczyk E, Fishman MC. A long purine-pyrimidine homopolymer acts as a transriptional diode. J Biol Chem 1995; 270:1791-1797.

70. Campuzano V, Montermini L, Lutz Y et al. Frataxin is reduced in Friedreich ataxia patients and is associated with mitochondrial membranes. Hum Mol Genet 1997; 6:1771-1780.

71. Cossée M, Campuzano V, Koutnikova H et al. Frataxin fracas. Nat Genet 1997; 15:337-338.

72. Monros E, Moltó MD, Martinez F et al. Phenotype correlation and intergenerational dynamics of the Friedreich ataxia GAA trinucleotide repeat. Am J Hum Genet 1997; 61:101-110.

73. Lamont PJ, Davis MB, Wood NW. Identification and sizing of the GAA trinucleotide repeat expansion of Friedreich ataxia in 56 patients—Clinical and genetic correlates. Brain 1997; 120:673-680.

74. Schols L, Amoiridis G, Przuntek H, Frank G, Epplen JT, Epplen C. Friedreich ataxia. Revision of the phenotype according to molecular genetics. Brain 1997; 120:2131-2140.

75. Bidichandani SI, Ashizawa T, Patel PI. Atypical Friedreich ataxia caused by compound heterozygosity for a novel missense mutation and the GAA triplet-repeat expansion. Am J Hum Genet 1997; 60:1251-1256.

76. Cossée M, Puccio H, Gansmuller A et al. Inactivation of the Friedreich ataxia mouse gene leads to early embryonic lethality without iron accumulation. Hum Mol Genet 2000; 9:1219-1226.

77. De Michele G, Filla A, Cavalcanti F et al. Atypical Friedreich ataxia phenotype associated with a novel missense mutation in the X25 gene. Neurology 2000; 54:496-499.

78. Babcock M, de Silva D, Oaks R et al. Regulation of mitochondrial iron accumulation by YFH1, a putative homolog of frataxin. Science 1997; 276:1709-1712.

79. Branda SS, Cavadini P, Adamec J, Kalousek F, Taroni F, Isaya,G. Yeast and human frataxin are processed to mature form in two sequential steps by the mitochondrial processing peptidase. J Biol Chem 1999; 274:22763-22769.

80. Priller J, Scherzer CR, Faber PW, MacDonald ME, Young AB. Frataxin gene of Friedreich's ataxia is targeted to mitochondria. Ann Neurol 1997; 42:265-269.

81. Gordon DM, Shi Q, Dancis A, Pain D. Maturation of frataxin within mammalian and yeast mitochondria: one-step processing by matrix processing peptidase. Hum Mol Genet 1999; 8:2255-2262.

82. Knight SA, Sepuri NB, Pain D, Dancis A. Mt-Hsp70 homolog, Ssc2p, required for maturation of yeast frataxin and mitochondrial iron homeostasis. J Biol Chem 1998; 273:18389-18393.

83. Wilson RB, Roof DM. Respiratory deficiency due to loss of mitochondrial DNA in yeast lacking the frataxin homologue. Nature Genet 1997; 16:352-357.

84. Radisky DC, Babcock MC, Kaplan J. The yeast frataxin homologue mediates mitochondrial iron efflux. Evidence for a mitochondrial iron cycle. J Biol Chem 1999; 274:4497-4499.

85. Rötig A, deLonlay P, Chretien D et al. Frataxin gene expansion causes aconitase and mito-chondrial iron-sulfur protein deficiency in Friedreich ataxia. Nature Genet 1997; 17:215-217.
86. Gardner PR, Rainieri I, Epstein LB, White CW. Superoxide radical and iron modulate aconitase activity in mammalian cells. J Biol Chem 1995; 270:13399-13405.
87. Lill R, Diekert K, Kaut A et al. The essential role of mitochondria in the biogenesis of cellular iron-sulfur proteins. Biol Chem 1999; 380:1157-1166.
88. Foury F. Low iron concentration and aconitase deficiency in a yeast frataxin homologue deficient strain. FEBS Lett 1999; 456:281-284.
89. Isaya G, Adamec J, Rusnak F et al. Frataxin is an iron-storage protein. Am J Hum Genet 1999; 65(suppl.):A33 (abstract).
90. Musco G, Stier G, Kolmerer B, Adinolfi S, Martin S, Frenkiel T et al. Towards a structural understanding of Friedreich's ataxia: The solution structure of frataxin. Structure Fold Des 2000; 8:695-707.
91. Dhe-Paganon S, Shigeta R, Chi YI, Ristow M, Shoelson SE. Crystal structure of human frataxin. J Biol Chem 2000 Jul 18 [epub ahead of print].
92. Cho SJ, Lee MG, Yang JK, Lee JY, Song HK, Suh SW. Crystal structure of *Escherichia coli* CyaY protein reveals a previously unidentified fold for the evolutionarily conserved frataxin family. Proc Natl Acad Sci USA 2000; 97:8932-8937.
93. Lamarche JB, Côté M, Lemieux B. The cardiomyopathy of Friedreich ataxia morphological observations in 3 cases. Can J Neurol Sci 1980; 7:389-396.
94. Waldvogel D, van Gelderen P, Hallett M. Increased iron in the dentate nucleus of patients with Friedreich ataxia. Ann Neurol 1999; 46:123-125.
95. Delatycki MB, Camakaris J, Brooks H, Evans-Whipp T, Thorburn DR, Williamson R et al. Direct evidence that mitochondrial iron accumulation occurs in Friedreich ataxia. Ann Neurol 1999; 45:673-675.
96. Emond M, Lepage G, Vanasse M, Pandolfo M. Increased levels of plasma malondialdehyde in Friedreich ataxia. Neurol 2000; 55:1752-1753.
97. Wong A, Yang J, Cavadini P, Gellera C, Lonnerdal B, Taroni F, Cortopassi G. The Friedreich ataxia mutation confers cellular sensitivity to oxidant stress which is rescued by chelators of iron and calcium and inhibitors of apoptosis. Hum Mol Genet 1999; 8:425-430.
98. Ben Hamida M, Belal S, Sirugo G et al. Friedreich ataxia phenotype not linked to chromosome 9 and associated with selective autosomal recessive vitamin E deficiency in two inbred Tunisian families. Neurology 1993; 43:2179-2183.
99. Di Mascio P, Murphy ME, Sies H. Antioxidant defense systems: The role of carotenoids, tocopherols, and thiols. Am J Clin Nutr 1991; 53:194S-200S.
100. Lodi R, Cooper JM, Bradley JL, Manners D, Styles P, Taylor DJ et al. Deficit of in vivo mitochondrial ATP production in patients with Friedreich ataxia. Proc Natl Acad Sci USA 1999; 96:11492-11495.
101. Wilson RB, Lynch DR, Farmer JM, Brooks DG, Fischbeck KH. Increased serum transferrin receptor concentrations in Friedreich ataxia. Ann Neurol 2000; 47:659-661.
102. Hentze MW, Kühn LC. Molecular control of vertebrate iron metabolism: mRNA-based regulatory circuits operated by iron, nitric oxide, and oxidative stress. Proc Natl Acad Sci USA 1996; 93:8175-8182.
103. Santoro L, Perretti A, Lanzillo B et al. Influence of GAA expansion size and disease duration on central nervous system impairment in Friedreich's ataxia: contribution to the understanding of the pathophysiology of the disease. Clin Neurophysiol 2000; 111:1023-1030.
104. Puccio H, Simon D, Cossee M, Criqui-Filipe P, Tiziano F, Melki J et al. Mouse models for Friedreich ataxia exhibit cardiomyopathy, sensory nerve defect and Fe-S enzyme deficiency followed by intramitochondrial iron deposits. Nat Genet 2001; 27:181-618.
105. Rustin P, von Kleist-Retzow JC, Chantrel-Groussard K, Sidi D, Munnich A, Rotig A. Effect of idebenone on cardiomyopathy in Friedreich's ataxia: A preliminary study. Lancet 1999; 354:477-479.

INDEX